REVOLUTION IN PHYSICS

by
BAHRAM
KATIRAI

NOOR PUBLISHING COMPANY
THORNHILL, ONTARIO, CANADA

NOOR PUBLISHING COMPANY

World Wide Web: www.ipoline.com/~noor

E-Mail: katirai@ipoline.com

Copyright © 1993, Bahram Katirai

All Rights Reserved

THIS, OR PARTS THEREOF, MAY NOT BE REPRODUCED IN ANY FORM WITHOUT WRITTEN PERMISSION OF THE AUTHOR

Thornhill, Ontario, Canada

Canadian Cataloguing in Publication Data

Katirai, Bahram, 1948 -
    Revolution in Pysics.

Includes bibliographical references and index.
ISBN 0 - 9682359 - 0 - 5

1. Physics. I. Title.
QC 21 . 2 . K37 1997   530   C97- 931340 - 6

CONTENTS

Acknowledgment iv

Introduction v

History of Ether 1

Ether - The Building Block of all Physical Existence 13

Qualities of Ether 15

Nuclear Explosion vs. Non-Nuclear Explosion 17

Light and Sound 21

Simultaneous Light and Sound 23

Ether and the Magnetic Force 25

The Samareh (Electromoon) 31

Electron-Positron Pair Production 39

The Electron Wave 41

The Nucleus and Its Spin 45

Gravity and Einstein's Erroneous Postulates 49

Inertia 77

The Electric Field 79

Radiation as Waves of Ether 85

Unitary Field 91

The Transmission of Electrical Energy in a Conductor 95

The Motion of Electrons in Ether 97

Comets and Ether 99

Planetary Rings 109

The Motion of the Earth in Ether 121

The Motion of Asteroids in Ether 119

The Aurora Borealis 125

The Sun As A Source of Light 127

The Source of the Solar Wind 137

Pulsars 143

The Mysterious Explosion of 1908 146

The Earth Drags Ether 149

Stellar Aberration 153

The Michelson and Morley Experiment 161

Einstein's Rejection of Ether and Evidence of Fraud 169

Einstein's Relativity 198

Einstein's Ideas 248

Einstein and His Fame 252

Humanity's Loss: The Ether Propulsion System 263

Conclusion 269

Laws of Reflection of Light and Sound Waves 278

The Refraction of Light and Sound Waves 280

The Laws of Interference of Light and Sound Waves 281

The Independence of the Speed of a Wave from its Source .. 282

Laws of Doppler Effect for Light and Sound Waves 284

Photon theory: Evidence of Misrepresentation 285
Einstein's Misrepresentation of Maxwell's Theory: Evidence of Fraud.. 298
"The Relativity of Simultaneity"... 311
Longitudinal and Transverse Waves of Light................... 313
Mechanical View of Ether 316

Photon theory: Evidence of Misrepresentation 285
Einstein's Misrepresentation of Maxwell's Theory: Evidence of Fraud... 298
"The Relativity of Simultaneity"... 311
Longitudinal and Transverse Waves of Light................... 313
Mechanical View of Ether 316

Introduction

In this book, you will be introduced to a series of new scientific discoveries which may well be regarded in the future as the most important findings in physics, discoveries so significant that they will revolutionise our understanding of the physical world. This book makes the claim that space is not empty, but rather, is filled with an invisible substance called "ether". Scientists of the nineteenth century were very much convinced of ether's existence, yet belief in ether was abandoned at the beginning of the twentieth century. Proofs of its existence are so strong and clear that the author is confidant of its restoration as a concept of physics. This book provides compelling evidence by which the reader may judge for him or herself.

It will also be shown how the three forces of nature, namely electric, magnetic and gravitational, are created by currents in ether. You will be introduced to the discovery of a new subatomic particle, the "samareh", which revolves around an electron. The effect of the motion of this particle in ether solves many mysteries that have been troubling physicists since the middle of the nineteenth century, such as why an electron in motion creates a magnetic field in the "right-hand screw direction". As well, the samareh provides a very simple solution to the problem of the closed loop or poles of a magnet. It also explains the electron wave, electron diffraction, the positron, etc. It will be shown that the nucleus of an atom spins, and that its spin in ether creates the gravitational force. Problems about the force of gravity are solved with amazing simplicity.

This book will clarify how a disturbance in ether can generate light just as a disturbance in air can generate sound. Sparks, comet tails and planetary rings can be explained as disturbances in ether. Even sunlight can be understood as strong disturbances in ether rather than from hypothetical nuclear reactions that supposedly occur deep within the sun. The discoveries are many, ranging from the discovery of a new subatomic particle to discoveries about stars, comets, etc.

How did ether come to be discarded? The author, during many years of research, has uncovered a series of misrepresentations involving the concept of ether. Here you will find clear, irrefutable evidence exposing misrepresentation after misrepresentation committed in the name of science by a mathematician pretending to be a physicist. Evidence presented here will prove why the actions of this individual amount to the greatest fraud ever perpetrated in science. This book will shock the world into realising how one individual, by means of mathematical manipulation, distortion of facts and false information, was able to highjack the scientific community and mislead mankind into abandoning the truth about ether. As a consequence, he effectively slowed down the rapid progress of science. Without doubt, the damage this individual has caused to the world of science is beyond measure. If he had not misled humankind, and had not retarded the rapid progress of science, we would, by now, have explored many planets.

The truths contained in this book will awaken many to realise that what they have learned throughout their lives as facts are not necessarily true at all, but a mass of misrepresentations, distortions of truth and fictions presented in the name of modern physics. After reading this book, you will most probably change many of the ideas and views that you acquired in schools and universities as facts. The discoveries presented here are only the tip of the iceberg.

It is hoped that readers will carry out more research and investigation. It is hoped that these discoveries will not have been made in vain; that the reader will do something about them. Remember, if you do not arise to help, and instead choose to observe silence, then this book, like many others, will have no effect. Your children and your children's children, generation after generation, will continue to remain in darkness.

Remember also that whether you are a scientist or not, you can still help.

The author has made every effort to simplify concepts by avoiding difficult mathematics and terminology, so that everyone may see the truth without difficulty.

Chapter 1

History of Ether

After years of research and investigation, here is a brief account of the history of ether that is radically different from what is generally shown in famous scientific publications or text books. The nature of some of the information you are about to read is such that you might at first resist accepting it, thinking that it could not possibly be true. However, after reading the rest of the book you will notice that all claims have been substantiated with clear and irrefutable evidence, including over 250 footnotes.

For those who are not familiar with the concept of ether, here is a brief description.

Ether, according to the understanding of 19th century physicists, is a transparent, invisible substance made of extremely small particles filling all space, including the space occupied by ordinary matter. All atoms are made from ether particles. Just as atoms are the building blocks of all that exists physically, so ether particles are the building blocks of all the atoms. Just as air molecules fill the space around us, similarly invisible ether fills the space around us. The space within atoms or between atoms and molecules, and the space between planets and stars, is filled with invisible ether particles.

We know that sound is generated from the vibration of air. Similarly, light is generated from the vibration of ether. Gravity and the electric and magnetic forces are created from currents of ether and transmitted from one place to another. Just as the flow of air creates pressure or suction force, similarly the flow of ether creates all the magnetic forces of repulsion or attraction.

Ancient philosophers believed in an invisible substance which they called "aether" or "ether". Its constituent was thought to be very similar to the air we breath, but rarefied, that is, with much lower density.

In 1667, Robert Hooke suggested that light is the vibration of ether, and that light of different colours must be due to different rates of vibration. Later, in 1690, Christian Huygens developed Hooke's theory that light is the longitudinal wave of ether.

At about the same period, Sir Isaac Newton (1672) advocated the idea that light consists of streams of particles called "corpuscles". In 1752, Thomas Melville and Gaspard de Courtivron suggested that the colour of light was determined by the velocity of the corpuscles. Later, this was disproved by showing that the satellites of Jupiter do not change colour at the moment of an eclipse, proving that all colours travel at the same speed. (If colour were determined by speed, then following an eclipse of the sun the fastest-moving colour would appear first, followed by the other colours in succession, until finally the sun would resume its normal yellow. Since this is not the case, it must be assumed that all colours travel at the same speed.)

The discovery by James Bradley (1725) of the aberration of light[1] (the apparent displacement of the stars due to the earth's orbital motion), further disproved the corpuscular theory, because all stars share the same displacement, indicating that the velocity of light is independent of that of its source. This is easily understood if light is seen as a series of waves in a stationary ether, but it cannot be understood how, in the corpuscular theory, the corpuscles emitted by a moving body can fail to share in that motion.

---

[1] - *Philosophical Transactions of the Royal Society of London*, xxxv (1728), p. 637.

The conclusion of certain scientists of the time, such as Leonhard Euler, was that, "light is in the ether the same thing as sound in the air".

In 1800, Thomas Young found further evidence for the wave theory of light by the effect of films on light, a phenomena that also could not be explained by the corpuscular theory. He and later Augustin Fresnel (1815) developed the principle of interference, a principle which further proved, beyond the shadow of a doubt, that light is a wave, and not corpuscular.

In 1857, Gustav R. Kirchhoff[2] was able to prove that the velocity of an electric signal along a wire is equal to the velocity of light ($c = 3.1 \times 10^{10}$ cm/sec). Michael Faraday suggested that the ether which is the agent in the propagation of light might also be the agent in the transmission of electromagnetic signals.[3] Later in 1864, James Clerk Maxwell,[4] who was a supporter of Faraday's theories, was able to show that both light and the electric signals along conductors are transmitted by the medium of ether[5] and both could be generated by electrical or magnetic disturbances in ether. Maxwell showed that both light and electric signals could be explained by a single system of stresses in the ether. Maxwell suggested that electric and magnetic forces arise from the flow of ether.[6] He also suggested that the constituent of ether is not atoms or molecules as it is in gases or solids, and that therefore the law

---

[2] - *Phil. Mag.* xiii (1857), p. 393; *Ann. d. phys.* c (1857), p. 193.

[3] - Experimental Researches, p. 3075.

[4] - *Philosophical Transactions of the Royal Society of London*, Vol. 155, (1865), YMCCCLXV, p. 459.

[5] - ibid. P. 460, (4).

[6] - ibid. p. 460, (3).

of stresses in ether would be quite different from the law of stresses in gases or solids. Later, Heinrich R. Hertz, with the understanding that light is a wave of ether and that a disturbance in ether could create a wave, was able to prove that a sudden change in a magnetic field could generate a disturbance (radio wave) in ether, and that such a wave is propagated in space by the medium of ether.[7] He also showed that the laws concerning these waves are very similar to those of light and sound waves.[8] Hertz' discovery of the radio wave further confirmed the existence of ether.

George Stoke (1845), in trying to explain the reason for the aberration of light, came to the conclusion that the earth carries ether as it moves around the sun,[9] just as the atmosphere of air is carried. In 1871, A. A. Michelson[10] conducted an experiment to find out whether the motion of the earth through ether creates a wind of ether on the surface of the earth or whether the earth carries an envelope of ether. His experiments showed that near the surface of the earth there is no wind resulting from the motion of the earth through ether. Michelson concluded[11] that his work confirmed Stoke's conclusion that the earth carries, or "drags", ether. Heinrich Hertz (1890), suggested that all ponderable moving bodies completely drag ether. Later Michelson's experiments were carried out at different heights, such as that of a mountain, in hopes of finding the ether wind, yet all experiments showed negative results.

---

[7] - Miscellaneous Papers by Henrich Hertz, with an Introduction by Prof. Philipp Leonard, (1896), p. 319.

[8] - *Ann. d. Phys.* xxxiv (1888), p. 610; *Electric Waves* (English ed.), p. 137.

[9] - *Phil. Mag.* xxvii (1845), p. 9; xxviii (1846), p. 76; and xxix (1846), p. 6.

[10] - *Amer. Journ. Sci.* xxii (1881), p. 20.

[11] - *Amer. Journ. Sci.* xxi (1881), p. 386.

To explain why no ether wind could be found, Max Planck (1899) suggested that the earth carries with it an atmosphere of ether in the same way that it carries an atmosphere of air.[12] Since at the time there was no rocket system, and it was impossible to go far above the earth's surface to heights where different layers of ether are in relative motion, where the ether that the earth drags meets the ether in outer space or the solar wind, all experiments, including Michelson's, failed to find the wind.

FitzGerald[13] suggested that a solid object moving through stationary ether might contract in the direction of motion. In other words, a moving object encountering the resistive pressure of ether will become deformed in the direction of motion. In 1892, Hendrik Antoon Lorenz,[14] who had suggested that moving bodies do not drag ether, offered a different reason as to why Michelson could not find the ether wind. He postulated that since the earth moves in ether (without dragging ether) the resistive pressure of ether (the magnetic pressure) must compress the earth's atoms together in the direction of motion, making the length shorter. The amount of contraction of the earth might be just the necessary amount, only 6 centimeters, to explain why Michelson could not detect the wind.

Based on the assumption that the maximum speed a moving body can have is the speed of light, Lorenz developed the formula for relative length. According to Lorenz's formula, the faster a body moves, the shorter it becomes, and if the body moves with the speed of light, it

---

[12] - *Proc. Amst. Acad.* (English ed.), i (1899), p. 443.

[13] - FitzGerald's Scientific Writings, p. 557.

[14] - *Versl. K. Akad. W. Amsterdam.* 1, 74, 1892. See also English translation; Collected papers, H. A. Lorentz, Vol. IV, p. 219 and 221, The Hague, Martinus Nijhoff, 1937.

will have zero length. Later, Lorenz developed the formula for relative mass. A body moving in ether encounters a resistive pressure of ether, and as a result, behaves as if it has a greater mass for its acceleration. In 1900, Joseph Larmor predicted that a moving clock will run more slowly than when it is stationary, reasoning that when a clock is in motion, its moving parts will be slowed down by the pressure of ether. As a result, the clock will run more slowly. The amount that a clock slows down is so small that it could be confirmed only by the aid of an atomic clock.

In 1905, Henri Poincaré developed the general principle of relativity with ideas such as simultaneity. A few months later, Albert Einstein presented a paper[15] which had the same mathematical structure and the same relativity formulas as those of Lorenz, along with Poincaré's ideas. Einstein's paper was so ambiguous that even decades later the media and scientific press claimed that only two people had understood it. Many scientists claimed that Einstein's paper is a copy of Lorenz and Poincaré's papers. Although the formulas and their interpretations were exactly the same as those of Lorenz and Poincaré's who had published them before Einstein,[16] yet Einstein, in the books he wrote to explain his relativity paper, claimed that the cause of relativity is the speed of light. According to Einstein, relativity has nothing to do with ether, but rather with the speed of light and how the observer sees. Einstein claimed that contraction of length, a moving clock running slow, and mass increase due to velocity, are effects of the speed of light,

---

[15] - *Ann. d. Phys.* xvii (Sept. 1905), p. 891. See Appendix for a copy of the English translation of the relativity paper called "Electrodynamics of Moving Bodies" by A. Einstein.

[16] - In later chapters all the references and evidences are given.

not ether. Einstein, with these interpretations, removed ether from relativity.[17]

Many scientists objected to Einstein's relativity theory on the basis that it is sheer nonsense to claim that the rate a clock ticks, the contraction of length of a body, or a mass increase, could be caused by the speed of light, for there could be no cause and effect relationship between the speed of light and these phenomena. Many scientists realised that Einstein's paper was a misrepresentation, but were powerless to do anything, because Einstein's aid, Hermann Minkowski, introduced the idea of the fourth dimension and time travel, which captured the imagination of the masses at large by making all things seem possible, and turning everything in Einstein's favour.

Einstein also claimed that relativity disproves the existence of ether and therefore we must forget ether.[18]

In support of his relativity ideas that there is no ether, Einstein wrote another paper the same year, reviving Newton's old corpuscular theory of light, a theory which had already been disproved. Einstein claimed that light is not a wave of ether, but corpuscles or photons that move in empty space. To support his idea, he took Planck's formula, changed its form and called it the "photoelectric effect formula", which he claimed supported the particle nature of light. Since a particle cannot have a wavelength, he also fabricated an assumption that the different photons must have variant energy and that the energy of a photon must be proportional to the frequency of light. He offered no supportive reasons for his assumptions. Despite the fact that

---

[17] - *Van Nostrand's Scientific Encyclopedia*, sixth edition, Van Nostrand Reinhold Company, New York p. 2428.

[18] - Albert Einstein and Leopold Infeld, *The Evolution of Physics*, Simon and Schuster, New York, (1938), p.184-185.

his ideas had no scientific basis whatsoever, they were publicized as the greatest scientific works, and as a result won general acceptance. Consider, for example, the experiment normally used to illustrate wave interference, in which light is shone through two narrow slits and forms a curious pattern of light and dark bars on the wall opposite. Despite the fact that this phenomenon cannot possibly be explained by any action of particles, since a particle must go through one slit or the other, and there cannot be any question of the two slits simultaneously affecting its motion, nevertheless Einstein's papers changed the direction of the whole science of physics.

It will be explained later that the phenomena and experiments which, up to that time, were considered to be proofs of ether, were interpreted by Einstein in such a way as to disprove its existence. For example, the relativity theory which had been originally founded on the basis of ether suddenly became 'proof' of ether's non-existence. The relativity formulas originally belonging to Lorenz became Einstein's formulas. The mass increase due to velocity which was a proof of the effect of ether and supported Lorenz's relativity formulas became proof of Einstein's ideas and was used against ether. The photoelectric effect phenomenon, which actually supports the wave nature of light, was misrepresented and used as a proof that light is composed of corpuscles, which negates the need for ether.

Einstein also misrepresented Maxwell's electromagnetic theory of light, claiming that it had nothing to do with ether. Einstein claimed that Maxwell developed equations that did not incorporate the existence of ether and only incorporated the electric and magnetic fields. He evaded the fact that Maxwell developed his theory on the basis that these fields are created by the current of ether.

Einstein also misrepresented Hertz' discovery of the radio wave which was based on ether, in such a way as to exclude ether, making it support his own ideas. The radio wave, which was a proof of ether, he used against ether.

Einstein also claimed that electromagnetism and mechanics are different and that there is no way of explaining the one in terms of the other. In fact, both are ether effects, which is clearly explained in the following chapters. The phenomenon of stellar aberration which, up to Einstein's time, was considered to be an indication of the earth's drag of ether, became a proof against ether. Michelson's experiment indicating the earth's drag of ether was misinterpreted by Einstein as proof against ether. Up to the present, relativity supporters claim that since, at the height of a mountain or balloon, very accurate experiments have failed to register the ether wind, this proves there is no ether at all. The fact has been evaded that in the middle of the 20th century, scientists were able to conduct experiments beyond the atmosphere, deep into outer space, where they indeed discovered very fast etheric (magnetic) winds which they called the "solar wind". They also discovered that this wind, which moves with a velocity of 450 kilometres per second, is continuously being pushed to the sides at a distance of several thousand kilometres from the earth by the bow shock (see page 93) and its invisible force, which acts as an umbrella or shield to protect the earth, the atmosphere and even the earth's magnetic field from the direct impact of the solar wind. This discovery explains why the Michelson experiment could not find the wind. A force that can shield the atmosphere of the earth from the fast moving (450 km/sec) solar wind can also shield the atmosphere from the slow moving (30 km/sec) wind arising from the motion of the earth through ether. These discoveries support Hertz's, Stoke's and Planck's theories that the earth drags ether.[19]

---

[19] - For further information see Chapters 28 and 30, " The Earth's Drag of Ether" and "Michelson's Experiment".

However, relativity supporters have evaded these discoveries and promoted Einstein's ideas.

Einstein also created the general misconception that a proof of relativity is a proof against ether. In many publications, ether, the existence of which had been repeatedly confirmed, was termed "hypothetical", and Einstein's ideas became "theories". Finally, the understanding that light is the vibration of ether was replaced by two opposite and contradictory theories of Einstein: namely, the photon theory and a new version of the electromagnetic theory. Relativity supporters claim that, in all cases, light behaves as a wave except in the two cases where Einstein and his supporters give opinions, in which case light behaves as a particle. The wave aspect of the theory was purported to be Maxwell's theory, but in reality it misrepresented it,[20] because the basis of Maxwell's theory was ether. Maxwell believed that light is an electromagnetic disturbance in ether. Einstein removed the ether from Maxwell's theory and claimed that light is an electromagnetic wave that propagates in empty space.

Einstein also discarded the simple explanation that electric and magnetic forces arise from the flow of ether, replacing it by a mathematical abstraction. According to Einstein, the only thing we need to know about a magnetic force is its amount and direction which can be represented geometrically as a field. We do not need to know the nature of the force, or how it can make itself felt at a distance.

In brief, through this kind of reasoning, Einstein was able to convince scientists that there is no need for ether. The understanding that electrical and magnetic forces are created by currents of ether was replaced by his dead-end proposal

---

[20] - See Chapter 31, "Einstein's Rejection of Ether" in the section called "Einstein's Misrepresentation of Maxwell's Theory: Evidence of Fraud". For more information see Appendix.

that the nature of these forces is unknown, since scientists, on the basis of Einstein's ideas, could find no answer for why magnetic or electrical forces are able to attract or repel from a distance. Einstein also misled mankind into believing that no measurable direct mechanical effect of ether has ever been detected, evading or unaware of the fact that magnets and electricity can do mechanical work and that magnetic and electrical forces are actually the effects of the current of ether.

Chapter 2

Ether - The Building Block of all Physical Existence

As the origin of all numbers is one, so all the infinite diversities of physical creation are made from one basic ingredient called the "ethereal matter" or "ether". If we examine nature, we see that there are infinite, diverse physical creations existing all around us. There is no end to the number and variety of shapes, sizes, colours, qualities, etc., that we can see. But, when we look into the chemistry of these physical creations, we see that they are all composed of a limited number of elements such as hydrogen, oxygen, and carbon. These elements themselves are made up of much simpler and more elementary particles, namely, protons, electrons, etc. In other words, looking into the atoms of all the elements, we see that the infinite varieties of existence are composed of only several basic ingredients. This analysis leads us to the question: could it be that these several ingredients themselves are composed of only one simple ingredient?

The fact that Marie Curie was able to split the nucleus of an atom shows that the nucleus of an atom is divisible and must therefore be made of smaller particles. Furthermore, the radioactive disintegration or decay which takes place over a very long period of time suggests that the nuclei of atoms are made of much smaller and simpler particles. These very small elementary particles composing the nucleus are those of ether. Not only are they extremely small, but also they are invisible, and they exist in alll of so-called 'empty' space. All atoms and molecules are made up of ether particles; they exist everywhere. They exist in the space inside an atom, in the space outside the atoms, in air, in outer space.

In brief, as atoms are the building blocks of all existence, ether is the building block of all atoms.

Chapter 3

Qualities of Ether

Ether, as the building block of all matter, leads us to the understanding that it must possess all the attributes of matter. In other words, what exists in the branches must also exist in the root. So all the natural, universal laws applicable to matter must also be applicable to ether. For example, ether must have density; it must have weight; it must be attracted by the force of gravity; it must flow; it must have viscosity; it must be subject to centrifugal force; it must vibrate, move, etc.

Since the particles of ether are extremely small, not only are they invisible but their motion through our body goes unnoticed. That is why, unlike the pressure of air that can be felt on the surface of our skin, the magnetic force, which is the pressure of the wind of ether, cannot be sensed on the surface of our skin, as ether particles can easily pass through our body.

If a certain volume of ether is compressed, the ether is elastic, just like the air in a balloon; that is to say, its volume is shrinkable. However, there is much difficulty in compressing ether, because the ether particles are so small that, unlike the molecules of air that cannot pass through the walls of a balloon, ether particles can pass through anything. They can pass through a wall of steel, no matter how thick it is.

So far, it has been mentioned that, as atoms are the building blocks of physical existence, so ether particles are the building blocks of all atoms. As the air exists around us in an invisible form, so ether exists around us in an invisible form. In the same space, there exists both air and ether particles.

The existence of ether as an invisible substance in space and as the building block of all physical existence is similar to water molecules, which exist in different forms in nature, such as humidity in air, water in the sea, or solids like ice or snow on the ground. These are the same substance in different forms: one fills the air, the other covers the ground; one is invisible, the other visible; one hard as rock, the other cannot be felt at all. All these substances are made from the same ingredient but exist in different forms. A piece of ice that weighs one kilogram occupies about 10 cubic centimetres, whereas the same amount of water molecules as vapour occupies more than 1600 times that volume. Similarly, ether particles as an invisible form in space occupy many times more volume than when they are a visible solid (matter).

In order to know more about the qualities of ether, it is helpful to investigate the difference between a nuclear explosion and a non-nuclear explosion, such as that of dynamite.

Chapter 4

Nuclear Explosion vs. Non-Nuclear Explosion

Scientists know that the nucleus of a radioactive element such as that of radium decays or disintegrates continuously. Over a long period of time, thousands or even millions of years, it loses much of its mass. Unlike this slow, natural decay, a nuclear explosion occurs suddenly and causes a great deal of disintegration in a very short period of time.

By bombarding uranium with neutrons, a chemist was able to split the nucleus of an atom into barium, lanthanum and cerium. Later it was discovered that by firing neutrons into the nucleus of a uranium atom, the nucleus, like a mini bomb, explodes. Some of the mass of the nucleus splits into fragments (fission products) and some vaporises to create magnetic pressure, shock waves and radiation, including light and heat, with tremendous energy. Scientists examining fission products realized that the mass of the original uranium atom exceeds the sum of the masses of the fission products. In other words, it was found that some of the mass disappears completely. Einstein and his relativity supporters claimed that by splitting the nucleus, some of the mass of the nucleus converts into light and heat energy.

The fact is that the break-up of the uranium atom into fragments must include vaporisation and release of ether. That is why it explodes. To think that some of the mass converts into light energy is like thinking that when dynamite explodes, some of the mass converts into sound energy. When the nuclei of the uranium atoms are broken up, some of the mass completely disintegrates into its elementary ether particles. That is why some of the mass disappears or cannot be collected and weighed. The vaporisation, the release of ether, and the sudden expansion of volume creates powerful and destructive shock waves.

A nuclear explosion is very similar to a non-nuclear explosion. When a dynamite bomb suddenly vaporises, that is to say, it explodes, a large amount of gas is released in a very short period of time. Since the volume of gas is much larger than the volume of dynamite as a solid, this sudden expansion of volume creates a very powerful and destructive sound wave in the air. As a solid, the molecules of dynamite are very densely packed, whereas in a gaseous state, they are widely separated. Similarly, in a nuclear explosion, there is a sudden expansion of volume as a result of vaporisation of some part of the nucleus into ether. In a nucleus the ether particles are very close to each other, whereas in space the ether particles are widely separated from each other. The sudden vaporisation and release of ether particles from a large number of atoms causes a sudden expansion of volume in ether that creates a very powerful and destructive ether shock wave moving at the speed of light. These shock waves move 900,000 times faster than sound. This is because the velocity of light is 300,000 kilometres per second, whereas the velocity of sound in air is only .333 kilometres per second. For this reason, the energy that the ether wave carries has a correspondingly greater effect than sound waves, and as a result, the destructive power is also 900,000 times greater.

Another way of looking at it is to consider that in the explosion of dymamite, only altoms and molecules separate, whereas in a nuclear explosion atoms disintegrate within themselves, that is to say, the nuclei disintegrate. That is why the distructive power is many times greater.

In a non-nuclear explosion, the cause of destruction is a gas shock wave, whereas in a nuclear explosion, the cause of destruction is mainly an etheric shock wave. A non-nuclear explosion, such as dynamite, is accompanied by a loud sound wave, and a nuclear explosion is accompanied by an intense wave and vibration of ether that we call "radiation".

Since the molecules of gas and ether are both present in a given explosion, every non-nuclear explosion is accompanied by a small amount of radiation such as light and heat; and a nuclear explosion is also accompanied by some sound waves.

In both cases, if the explosion occurs on the ground, a similar shape is created, that of the mushroom cloud. However, the size of one mushroom cloud is much larger than the other. Figure 1 shows the mushroom cloud made by a nuclear bomb. Figure 2 shows the mushroom cloud of a non- nuclear bomb.

Fig. 1

Fig. 2

Chapter 5

Light and Sound

One of the most important correlations between ether and air is that from the wave and vibration of air, sound is produced, whereas from the wave and vibration of ether, light is produced. As a sudden disturbance in air creates a sound wave, similarly a sudden disturbance in ether creates a light wave. That is why the sudden impact of a sledge hammer on a hard stone creates disturbances in both air and ether, and generates both light and sound. That is why many phenomena which generate sound can also generate light.[21] In the case of air, the human ear is only sensitive to air vibrations, and in the case of ether, the nerves of our eyes are only sensitive to ether vibrations. All the different melodious sounds are in reality made from a combination of different frequencies of air vibrations, and all the beautiful colours are made from the combination of different frequencies of ether vibrations. As ears can distinguish all the different voices and sounds in nature, so also can our eyes distinguish different people and objects by their light. As our ears are only sensitive to certain vibrations ranging from 20 to 20,000 cycles per second, so also are our eyes only sensitive to certain vibrations that range from 10^{14} to 10^{15} cycles per second. In both cases, sound and light, the higher the frequency, the greater the energy it carries and the more its penetrative power.

The following illustrates some other parallels between light and sound. For details, see the appendices.

---

[21] - For detail see "Simultaneous Light and Sound", p. .

1. The laws of the reflection of light from a surface are the same as for those of sound waves (Appendix A).
2. The laws of the refraction of light are the same as for those of sound waves (Appendix B).
3. The laws of interference of light waves are the same as for those of sound waves (Appendix C).
4. The laws of the independence of the speed of light from its source are the same as for those of sound waves (Appendix D).
5. The laws of the Doppler effect for light are the same as for those of sound waves (Appendix E).

Chapter 6

Simultaneous Light and Sound

Since the space around us is filled with both air and ether, many disturbances affect both air and ether and thus generate both sound and light simultaneously. The following are some examples of such phenomena, demonstrating that light and sound are indeed waves of ether and air respectively. Each creates both light and sound, light being, in some cases, a spark:

1. natural lightning;
2. collision of two hard surfaces (such as a sledgehammer's impact on hard stone creating a spark [light] and sound);
3. non-nuclear explosion (such as dynamite);
4. nuclear explosion;
5. connecting two live electrical wires;
6. very high energy electron hitting a metal plate (creates sound as well as light and x-rays, which have the same nature as light, except with a higher frequency);
7. fire;
8. corona discharge;
9. sparks in open air;
10. hot wire such as heating elements (audible at very close range);
11. friction (such as lighter flint).

Chapter 7

Ether and the Magnetic Force

Before Einstein, 19th century scientists such as Michael Faraday, James Clerk Maxwell and Heinrich Hertz had concluded that magnetic or electric forces arise from the current of ether.[22] However, belief in ether having been banished by Einstein, science of the 20th century is in the dark when it comes to explaining the basic forces of nature. Questions such as: - what is the magnetic force? - what is the electric force? - how and why are these forces able to act at a distance? - have remained unanswered. For this reason, these basic questions are avoided altogether. Scientists of the 19th century had a very good understanding of why the magnetic and electric forces are able to attract or repel from a distance.[23] However, Einstein changed all that. According to Einstein, a magnet creates "something" around it. Einstein claims that this "something", for unknown reasons, is able to attract a piece of iron from a distance. Here are his own words:

> *If, for instance, a magnet attracts a piece of iron, we cannot be content to regard this as meaning that the magnet acts directly on the iron through the intermediate empty space, but we are constrained to imagine ....- that the magnet always calls into being something physically real in the space around it, that something being what we call a "magnetic*

---

[22] - See Maxwell's Molecular theory published in the Philosophical Magazine, 1861-1862, 4th ser. vols 21 and 23. See also Maxwell's treatise on *Electricity and Magnetism*.

[23] - Encyclopedia Britannica, (1890), vol.XV, p. 276.

field." In its turn this magnetic field operates on the piece of iron, so that the latter strives to move towards the magnet. We shall not discuss here the justification for this incidental conception, which is indeed a somewhat arbitrary one.[24]

Einstein obviously feels uncomfortable with the notion that there is "something physically real" (i..e., ether) between the magnet and the iron, because he refuses to discuss it; and in fact dismisses this "something" as an "arbitrary" and "incidental conception". He cannot explain magnetism without it, however, so magnetism remains an unknown, unexplained phenomenon.

According to Einstein, the wording "electric field" and "magnetic field" sufficiently describes a force. He claimed that all we need to know about a force mathematically and geometrically is its amount and direction;[25] there is no need to know the nature of the force, or how it can make itself felt at a distance. Whether the force is from a horse, a car, a suction hose or a magnet, is irrelevant, so long as we know its amount and direction. By this kind of reasoning, Einstein created the belief that there is no need for ether. So far, his ideas have not been able to provide reasons for how these forces are able to act at a distance, or what their nature is. In the 19th century, scientists were able to explain these forces by the effect of the motion of ether, which provided the answer with simplicity

---

[24] - Albert Einstein, *Relativity, the Special & the General Theory*, Translated by R. W. Lawson, 3rd ed, Methuen & Co. Ltd. London, (1920), p. 63

[25] - Ibid.

and consistency.[26] In 20th century, however, in the name of modern physics, through mathematical manipulation and distortion of facts, Einstein removed the basic ingredient of ether and created many problems that no scientist has been able to solve ever since.

Here, in this and the next few chapters, it will be illustrated how the current of ether manifests itself as the magnetic force. Later on, it will be shown how the electric and gravitational forces are also created by the current of ether.

The magnetic phenomenon results from the flow of ether. Earlier, we saw parallels between air and ether, and here is another: just as from the current of air a force is created, similarly from the current of ether a force is created which we call the "magnetic force". Everyone knows that the force of a strong wind is created by a current of air. The faster the current of air, the stronger will be its force. If you, on a windy day, put your hand out the window, you will feel the force of the wind. The wind applies a pressure on your hand. We know that the current of air creates a force in the direction of its motion. In other words, the direction of the force is the same as the direction of the current.

Fig. 1

The flow of air from A to B creates a force in the same direction. The higher the speed and density of the air, the greater will be its force. Similarly, from the flow of ether from

---

[26] - For example, see H. A. Lorentz in "Electric and magnetic forces act by means of the intervention of the ether", Collected Papers, H. A. Lorentz, Vol. IV, Page 221, The Hague, Martinus Nijhoff, 1937.

one place to another, a force is created which is called the magnetic force. When a series of ether particles move from one place to another, all in one line, it creates a "line of force". The direction of the current of ether is the same as the direction of the force. The higher the speed and density of ether, the stronger will be its force.

In the case of the current of air, if a hand is raised against the current, the force of air is felt on the surface of the skin. However, in the case of ether, we do not feel its force because the ether particles are so small that they pass through the atoms and molecules of our hand, going in from one side and coming out the other, without applying any noticeable pressure.

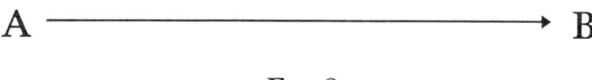

Fig .2

A line of force. The flow of ether from A to B creates a magnetic force in the same line and direction. The higher the speed and density of ether, the greater is its magnetic force.

The flow of air on the surface of the earth, as everyone knows, is called "wind". Similarly, the flow of ether on the surface of the earth is called "ether wind" or "magnetic wind".

In the case of air, if the wind is strong, it is called a "storm". Similarly, a strong wind of ether is called a "magnetic storm". A nuclear explosion creates both air wind and ether wind. After any nuclear explosion inside the atmosphere, not only is a strong wind of air created, but also a strong magnetic storm. The strong wind of air occurring after a nuclear explosion is the result of a sudden expansion of volume in air. The magnetic storm occurring after a nuclear explosion is the result of a sudden expansion of volume in ether which creates enormous pressure and ether wind. The velocity of this wind is much less than the velocity of light. It is similar to the speed of an air wind, which is much less than the speed of sound. A very strong wind of air moves at 150 kilometres per hour, while the speed of sound is approximately 1200 kilometres per hour. The speed of a magnetic storm in comparison with the speed of light is very slow. That is why the magnetic storms generated by a nuclear explosion are noticed after seeing the bright light of the explosion. Similarly, after a nuclear explosion the wind of air occurs long after hearing the sound of the explosion.

It will be proven later that magnetic storms caused by solar flares are also storms of ether. Since the storm moves much more slowly than light, the magnetic storms on earth are registered long after seeing the flare.

The earth has its own magnetic field and acts as a magnet. This is due to the continuous flow of ether from one pole to the other. Later on it will be reasoned that the Michelson and Morley experiment actually measured the velocity of the slow wind of ether from pole to pole (see the chapter called "Michelson's Experiment").

In Chapter 12, "Gravity and Einstein's Erroneous Postulates", it will be shown how a current of ether creates the force of gravity and why gravitational forces behave differently from magnetic forces. In the next chapter it will be shown how the motion of a newly discovered particle called the 'samareh' creates a magnetic field.

Chapter 8

The Samareh (Electromoon)

Here I have made an important discovery that will not only solve many problems in physics but also provide a very clear proof of ether. The discovery of a new particle, hereafter called the "samareh", or "electromoon" is probably one of the most important discoveries of this century. It is by far one of the greatest additions to our knowledge of the atom.

A samareh is a particle that revolves around an electron. In the same way that an electron moves around the nucleus of an atom, so the electron itself is a dynamic centre around which the samareh (electromoon) circles. Comparing an atom with the solar system, the nucleus of an atom resembles the sun, and the electrons resemble the planets. As planets have moons, so do electrons have samarehs.[27] Fig. 1 shows the sun and the path of the Planet Earth and its moon around the sun. Fig. 2 shows a hydrogen atom and the path of an electron and its samareh around the nucleus. Notice the resemblance of the one to the other.

---

[27] - Just as some planets have more than one moon, some electrons may have two or more samarehs. Since some planets have no moons, some electrons may have no samarehs.

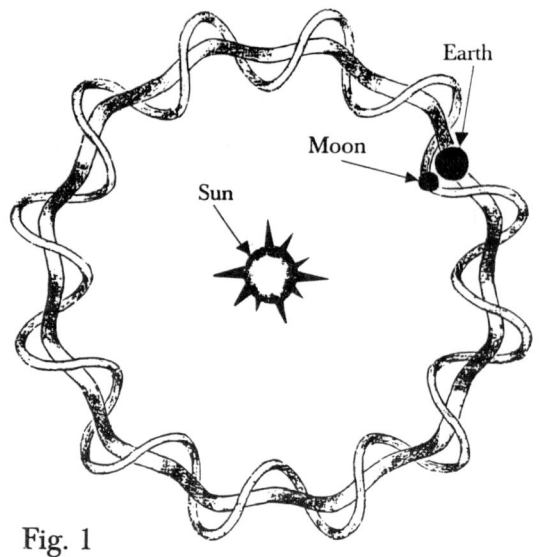

Fig. 1

Fig. 1 shows the sun and the path of the Planet Earth and its moon around the sun.

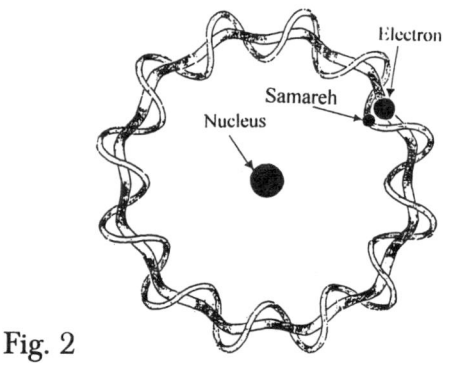

Fig. 2

Fig. 2 shows a hydrogen atom and the path of an electron and its moon (samareh) around the nucleus.

The discovery of the samareh solves with simplicity all the problems associated with electric and magnetic fields. For example, it explains the nature of magnetism, electric and magnetic fields, and why an electron in motion creates a magnetic field in the "right-hand screw" direction. It explains the reason for the existence of magnetic poles, north and south, the closed loop of a magnet, the electron wave, electron diffraction, electron-positron pair production, and all other phenomena associated with electricity and magnetism.

The effect of the revolution of the samareh in ether can be demonstrated by an example. Everyone knows that when you stir sugar into a cup of coffee, with only a few circular motions of the coffee stick, all the liquid inside the cup begins circulating along with the stick. Although the width of the stick is thin, its motion causes all the liquid inside the cup to circulate. Similarly, the circular motion of the samareh around the electron causes all the ether around the electron to circulate. In the above example, we see that a few revolutions of the stick per second is enough to cause all the liquid in the cup to circulate. However, the number of revolutions of the samareh per second is not a few but hundreds of billions. For this reason, the motion of the small samareh in a very short period of time can affect all the ether around the electron.

Fig. 3

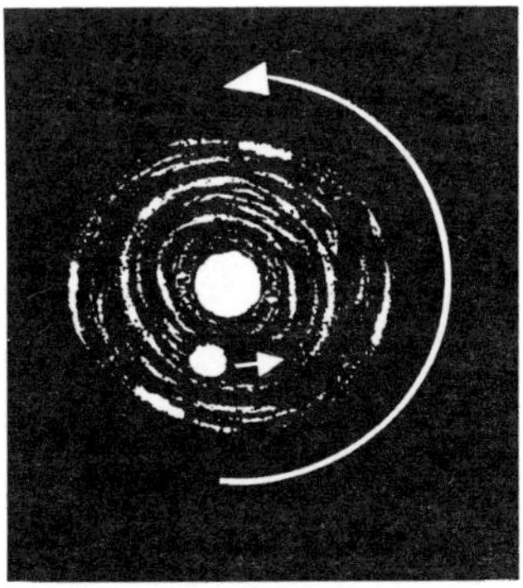

The figure shows a samareh circling around an electron. The circular motion of the samareh in ether causes the ether to circulate around the electron.

This circular motion of ether around the electron is called the magnetic field (Fig. 3).

Fig. 4 shows an electron moving in a straight line from A to B while the samareh spirals, circling the electron. The spiral motion of the samareh in ether generates a spiral current of ether in the right screw direction. The current of ether in this case is called the magnetic field, and its direction is known as the right-hand screw direction[28] of the field.

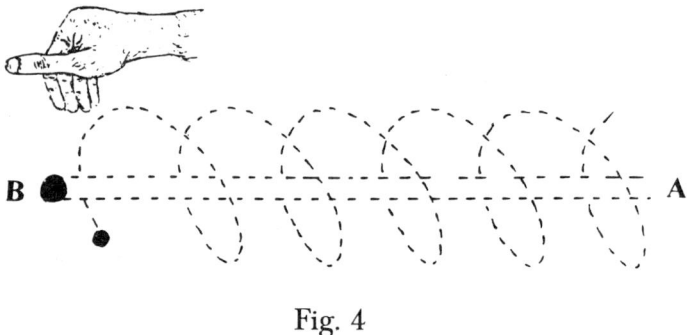

Fig. 4

Fig. 4 shows the path of an electron from A to B, showing the spiral path of the samareh. The effect of the spiral motion of the samareh in ether is to create a spiral current of ether (magnetic field) in the right-hand screw direction.

Since in an atom an electron revolves around the nucleus, the spiral motion of each samareh creates a loop current of ether. Fig. 5 shows an electron circling around a nucleus while a samareh is moving around the electron. The spiral

---

[28] - The lines of magnetic intensity around a current-carrying wire are circular, having the plane of the circle perpendicular to the axis of the wire. The direction of the line of force is determined by the "right-hand rule." When the wire is grasped by the right hand, the fingers encircling the wire, and the thumb pointing along the wire in the direction of current, the fingers encircle the wire in the direction of the lines of force.

motion of the samareh around the nucleus creates a loop current of ether, known as the magnetic field. Here we see how the spiral motion of the samareh creates a magnetic loop and makes of each atom a magnet.

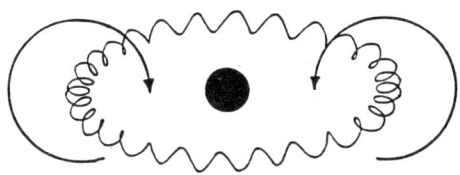

Fig. 5

Fig. 5 shows how an atom becomes a magnet. It shows the spiral path of a samareh around a nucleus The effect of the spiral motion of a samareh in ether is that it creates a spiral current of ether around the nucleus which is called a 'magnetic loop'.

Fig. 6 shows two atoms positioned top and bottom. The loop currents of ether of the two atoms are combined together and become a bigger loop.

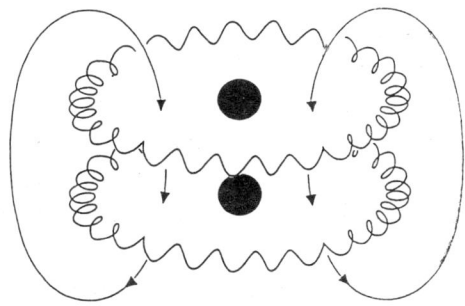

Fig. 6

Fig. 7 shows several atoms of a magnet. The atoms are positioned so that the ether current loops of all the atoms are combined together to make a much bigger loop that we call a magnet.

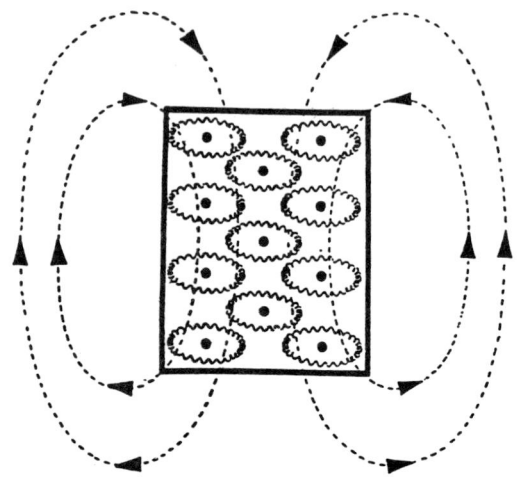

Fig. 7

This solution not only agrees with Ampere's explanation but also completes the missing puzzle. According to Ampere's theory of magnetism, the magnetic properties of a body arise from a multitude of tiny closed current loops within the body. In an un-magnetized body, the loops are oriented at random. The magnetising process consists of aligning these loops with the planes parallel to one another and their currents all circulating in the same direction.

Here we see how the motion of the samareh creates the current loops. Ampere had concluded that current loops exist but had not solved the mystery of what they were or how they were created.

The forces of attraction or repulsion between two magnets arise from the interaction between the loops. For example, two like magnetic poles repel each other because the flow of ether in both magnets is going in the opposite direction, disrupting each other's loop-currents and creating a pressure of ether between them (Fig. 8).

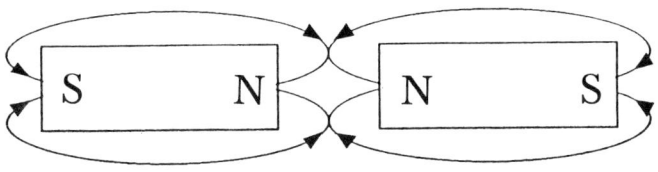

Fig. 8

The opposite poles of two magnets attract one another because the direction of the current of ether in both magnets is the same. As a result, they reinforce each other and cause the ether to escape from the space between the magnets, thus creating an attraction between them (Fig. 9).

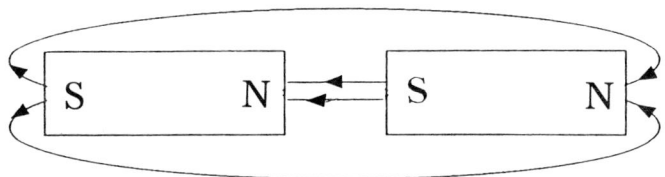

Fig. 9

It is interesting to note that the spiral motion of a samareh in an atom creates a loop, but because the electron circles around a nucleus, the loop of each atom also circles in the same direction as that of the electron.

Chapter 9

Electron-Positron Pair Production

So far, several different kinds of evidence have been presented which would point to the existence of the samareh. Here is another which may remove any further doubts you may have. In this chapter, it will be reasoned that cloud chamber photographs taken by physicists actually prove the existence of the samareh.

The orthodox followers of Einstein claim that cloud chamber experiments prove that when a beam of gamma rays passes through matter, an electron and a positron are created. They believe that the gamma ray, passing near a nucleus, transforms into an electron and a positron.[29] They claim; *"This is a phenomenon in which one can produce electrons out of a vacuum, that is, out of nothing."[30]*

The fact is that the high energy gamma rays cause the samareh (electromoon) to separate from its electron. This phenomenon is similar to that of the photoelectric effect phenomenon where light waves, falling on a metal plate, cause an electron to separate from its atom.

To think that gamma rays, which have no mass, can create "out of nothing" an electron and a positron, both of which do have mass, is absurd.

When we compare the diameter of the earth with that of the moon we see that the diameter of the moon is only a few times smaller than that of the earth. Similarly, the diameter of a samareh is not very different from that of an

---

[29] - W. A. Benjamin, *The Structure of Matter, An Introduction to Modern Physics*, USA, July 15, 1965), p. 281.

[30] - *The Realm of Science,* Touchstone Publishing Company, Louisville, Kentucky, V. 10, p.15.

electron. For this reason, scientists erroneously concluded that two equal-sized particles, namely an electron and a positron, are created. On the contrary, nothing is actually created; only the samareh and its electron[31] are separated from each other.

The separation of a samareh from an electron puts both particles into states of excitement and vibration. These vibrations in ether generate light or some other form of radiation which, reaching other electrons, may in turn separate the samarehs from their electrons. This explains the avalanche "shower" in which a little radiation causes an avalanche of positrons, electrons and radiation.

The orbiting of the samareh around the electron also solves a problem that has troubled the minds of many scientists as to why an electron travels in wave form. This subject will be discussed in the next chapter.

---

[31] - The author believes that the sizes of different electrons in an atom are not the same. If we assume that all electrons have the same size, it is as if we assume that all planets orbiting the sun have the same size. The emission of distinct series of line spectra such as the Balmer series may be attributed to differences in sizes of electrons.

Chapter 10

The Electron Wave

In this chapter you will find a very clear and irrefutable proof of ether.

In 1927, Davison and Germer, by sending a beam of electrons toward two slits, were able to produce interference similar to that of light waves. However, Einstein and his orthodox followers interpreted the results as evidence that electrons themselves are waves, and concluded that an electron must be both a wave and a particle at the same time. The author's understanding is that the movement of an electron in ether creates a shock wave in ether, a phenomenon similar to the movement of an object in water that creates a shock wave.[32] With a slow motion of one's finger immersed in water, in a bathtub one can see the shock wave, far ahead of the finger, moving on water. A beam of electrons carries many wave fronts, one after the other. Figure 1 shows how these wave fronts produce interference.

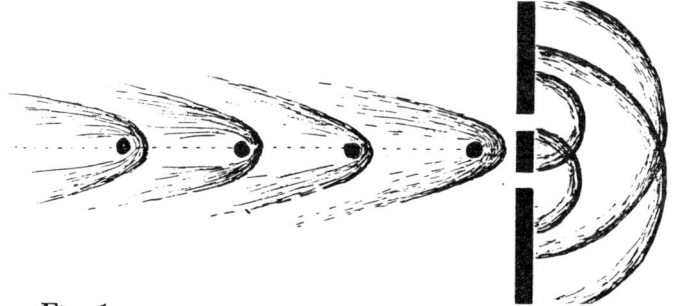

Fig. 1

Figure 1 shows a beam of electrons moving toward two slits and their wave fronts of ether producing interference.

---

[32] - It will later be shown that the movements of both comets and the earth also create shock waves in ether.

Furthermore the path of an electron is also in a wave form. Scientists have found no reason why electrons move in wave form. The discovery of the samareh solves this problem with a fascinating simplicity.

Fig. 2 shows the path of an electron around a nucleus.

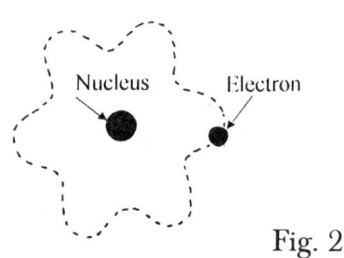

Fig. 2

According to the current understanding of scientists, the path of an electron around the nucleus is also in wave form

No one has been able to explain why an electron moves in this pattern. At one instance, it is approaching the nucleus, and at another, it is moving farther away from the nucleus. What is the force that causes the electron to behave in this manner? This puzzle has remained unsolved by

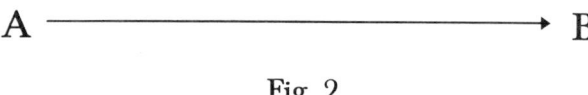

scientists. The discovery of the samareh answers very precisely the real cause of the electron wave.

To understand this concept, let us examine the orbital path of the earth around the sun. Due to the orbiting of the moon around the earth, the path of the earth around the sun is a spiral, because the gravitational force of the moon at different locations around the earth forces the earth to move spirally (Fig. 4).

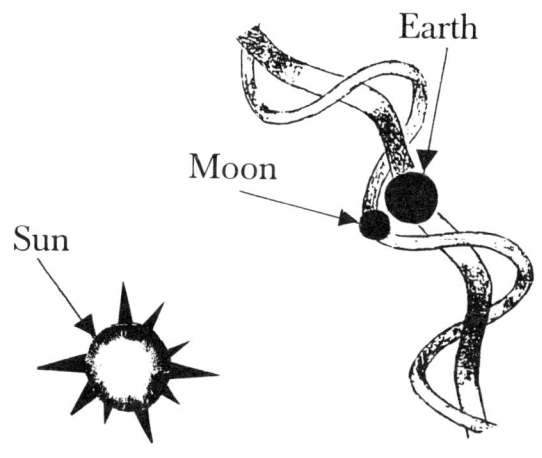

Fig. 4

The more we study the samareh, the more we find that there is indeed a striking mirroring of the atom with our solar system, proving the amazing symmetry in the creation of God. It proves there is more order to creation than we had thought.

Chapter 11

The Nucleus and Its Spin

The following, I believe, is another great scientific advancement of this century, the significance of which can only be measured in the future. There is a mirror image correlation between the sun and the nucleus of an atom. As the sun spins with a relatively high speed about its axis, the nucleus also has a high spin velocity about its axis.

There are numerous pieces of evidence to suggest that the nucleus is spinning, as the following examples indicate:

1) Radioactive material emits gamma rays. Scientists all agree that gamma rays come from the nucleus, that the nucleus vibrates and that gamma rays are the result of magnetic vibrations of the nucleus. They are, however, at a loss when it comes to explaining why the nucleus vibrates and why the vibrations create gamma rays. The most obvious explanation is that gamma rays are simply the vibration of ether. The spin of the nucleus creates a disturbance in the same way that sound is generated by the spin of the armature of an electric motor. The stability of the vibration in itself is a clear indication of the spin. This means that the frequency of gamma rays must be equal to the spin velocity of the nucleus.

2) A further evidence of spin is the orbiting of the electrons around the nucleus. The spin is the real cause of the orbital motion of the electron and the stability of these motions around the nucleus. Otherwise there is no reason why the electrons should not fall into the nucleus and the atoms crumble or collapse.

3) The spin of the nucleus explains why the nuclei of atoms have limited sizes, as spin determines size.[33] With a given spin velocity, if the diameter of a nucleus increases, there is then a point where the centrifugal force at the circumference will be larger than the cohesive force binding the particles of the nucleus together, thus causing the particles at the surface to be thrown off. This is the case for large nuclei such as that of the radium atom which gives off alpha particles.

4) In radioactive emissions of alpha and beta particles from a nucleus, it is found that the velocity of the emitted particles is very high, as though they were bullets being shot from a gun. How do the particles acquire such high initial velocity? A spinning nucleus could explain this phenomenon.

5) The following simple calculation is further evidence for the spin of the nucleus. If alpha particles gain their velocity from the spin of the nucleus, then the alpha particles must leave the circumference with a speed equal to the tangential velocity at the circumference. Furthermore, the number of spins per second must be equal to the frequency of gamma rays emitted by radium. Since the speed of alpha particles and the frequency of gamma rays are already known, we can calculate the circumference of the nucleus, which in turn can give the diameter of the nucleus of radium.

The frequency of gamma rays emanated by radium has been experimentally measured,[34] and found to be equal

---

[33] - The same law applies to the stars. The spin of the stars also explains why stars have limited sizes.

[34] - *College Physics*, By Francis Weston Sears, 3rd edition, Wesley Publishing Company Inc. London England, 1960, pp. 978 - 979.

to 10^{22} cycles per second. The number of spins of the nucleus must then be about 10^{22} revolutions per second.

The speed of alpha particles[35] emitted by radium is also known and is equal to 1.6×10^7 meters per second. The velocity of alpha particles ejected from the circumference of the nucleus must be equal to the circular velocity of the circumference (Fig. 1).

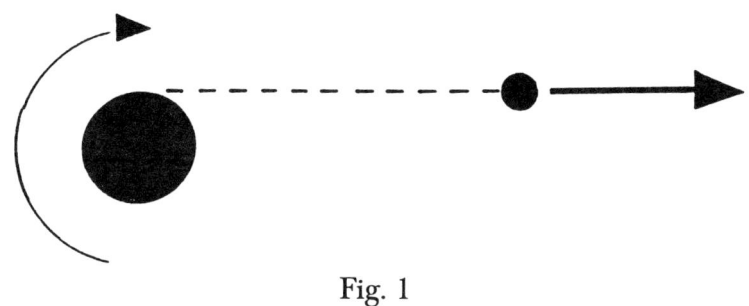

Fig. 1

Therefore, we can calculate the circumference of the nucleus of radium, which is:

$$\text{circumference} = \frac{\text{speed of alpha particles}}{\text{no. of revs. of nucleus per sec.}}$$

$$\text{or:} \quad = \frac{\text{speed of alpha particles}}{\text{frequency of gamma rays}}$$

---

[35] - Francis Weston Sears, *College Physics*, 3rd edition, Wesley Publishing Company Inc., London, England, (1960), p. 978-979.

$$= \frac{1.6 \times 10^7 \text{ m/sec.}}{10^{22} \text{cycles/sec.}}$$

$$= 1.6 \times 10^{-15} \text{ m}$$

Knowing the relationship of circumference to diameter to be:
Circumference = πr^2 = 3.14 (r^2) or 1.6×10^{-15} m = $3.14(r^2)$
Hence, the diameter of the nucleus of radium = $.5 \times 10^{-15}$ m. This result confirms the idea that Rutherford had about the size of the nucleus. According to Rutherford, the nucleus of an atom of gold is about 1/10,000 of the diameter of the entire atom. The findings here are not just an idea or estimation but accurate calculations which provide us with very important information about the nucleus. This shows that the size of the nucleus of an atom compared to the size of the entire atom is very small, as is the sun's diameter in comparison to the diameter of the solar system.

6) The spin of the nucleus solves all problems associated with gravity. It will be discussed in the next chapter.

Chapter 12

Gravity and Einstein's Erroneous Postulates

The force of gravity serves as further evidence to support the existence of ether. The simple reason that you find here for the existence of gravity solves many scientific problems.

The following are some basic questions about gravity:

1. What is the force of gravity and how is it created?
2. How is gravity able to attract at a distance?
3. Why is gravity different from the magnetic force?
4. Why does the force of gravity act independently from the magnetic force?
5. Why is the force all-attractive?
6. Why does the gravitational force not cause the nuclei of different atoms to crumble on each other?
7. Why is everything attracted towards the centre of the earth?
8. Why are the planets round?
9. Why do all objects positioned at the same height, regardless of their mass, fall at the same rate and have the same rate of acceleration in a vacuum?
10. Why, as the distance to the ground decreases, does gravity increase?
11. What is the reason behind Newton's inverse square law of gravity?

The existence of ether and the spin of the nucleus in ether together solve all problems about gravity with amazing simplicity. First of all, let us see how an atom creates its gravitational force. Here it will be illustrated that the current arising from the spin of the nucleus in ether creates the force of gravity.

The cause of gravity can be demonstrated by a simple experiment. When a spherical object spins very fast about an axis in water, at each of its poles a suction tunnel is created (Fig. 1). These suction tunnels serve as forces of attraction because through these tunnels there is a fast current of water towards the poles. At the poles the water is transferred spirally toward the equator and then from the sphere away into the surrounding area.

Fig. 1

This photo shows a spinning ball immersed in water. The photo shows only one of the poles of the spinning ball. Notice the suction line (gravity line) created at the pole. The other pole, not shown, was connected to a hand drill. The gravity line shown in the photo is not a straight line because the drill was shaking.

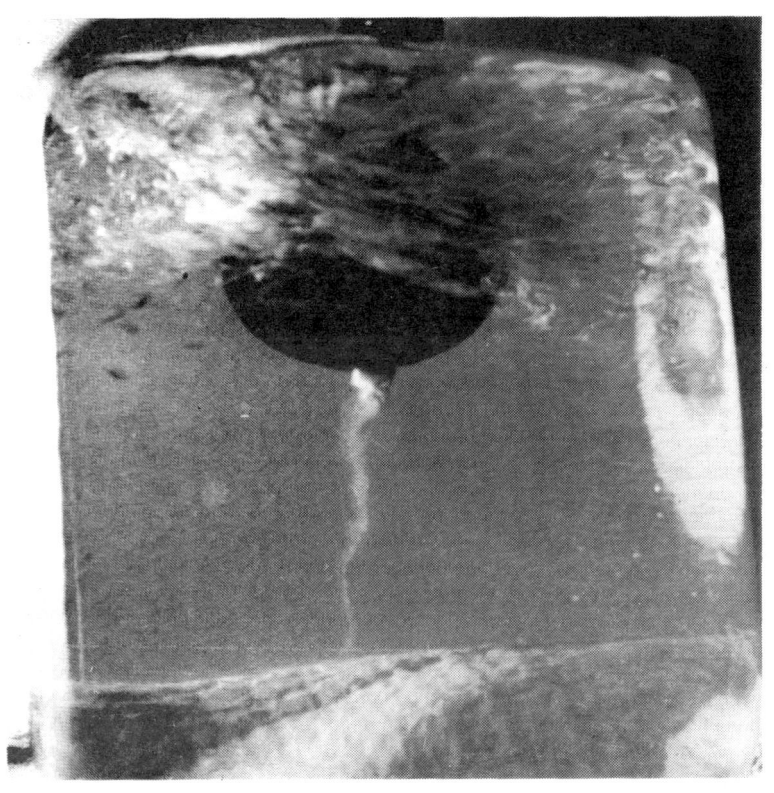

A spinning nucleus creates at its poles two suction lines called 'the gravitational force'.

Fig. 1A

This is because at the poles there is minimum motion, whereas at the equator, motion is at a maximum. Hence, at the equator there is maximum centrifugal force, while at the poles there is none. As a result, at the poles two suction tunnels are produced which continuously suck the water toward the sphere, which is then transferred towards the equator where the water is pushed away from the sphere into the surrounding area by the centrifugal force. Similarly, the spin of the nucleus in ether produces two suction tunnels at the poles of the nucleus, both of which continuously suck ether toward the poles. The ether is then transferred towards the equator where it is pushed away into the surrounding area by the centrifugal force. The faster the sphere spins, the longer and narrower the suction tunnels become. The force of suction at the poles is able to act at a much greater distance than the force of pressure at the equator. This is because at the poles, the ether passes through a very narrow tunnel and moves with a very high speed, while at the equator, the speed and pressure of the ether is lowest as it becomes scattered in a wide area covering the entire circumference. In the case of the spin of the nucleus in ether, pressure at the equator is dissipated and lost by the motion of electrons and their samarehs. Since the nucleus spins with a very high velocity, equalling hundreds of thousands of trillions of revolutions per second, the suction tunnels consequently are very long and narrow (Fig. 2).

Fig. 2

Looking at the figure, one can see why the force of gravity has some characteristics that are different from the magnetic force. The force of gravity is not a closed loop, while the magnetic force is a closed loop force. The forces of gravity (suctions) at both poles are attractive, whereas the magnetic forces at the poles are opposite, that is to say, repulsive and attractive.

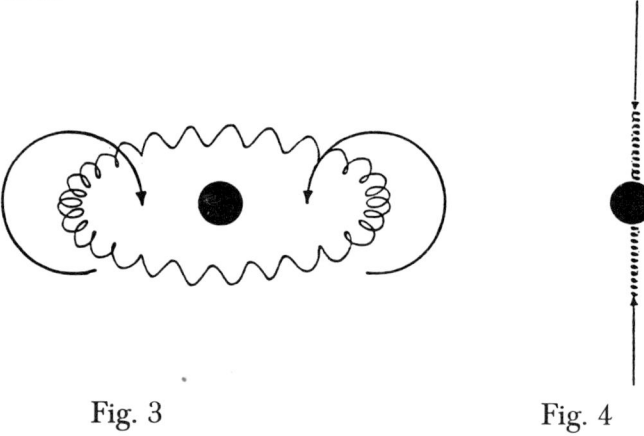

Fig. 3   Fig. 4

As to the question of why the gravitational force between two nuclei will not cause the nuclei of the atoms to combine and become one, here it will be reasoned that due to the spiral nature of the gravity lines and due to the fact that a nucleus is spinning, two nuclei do not combine to become one. To see the reason why, let us assume two nuclei are positioned close to each other. There are only three possibilities. The first possibility is that the suction line of one is directly locked onto the nucleus of another (Fig. 5).

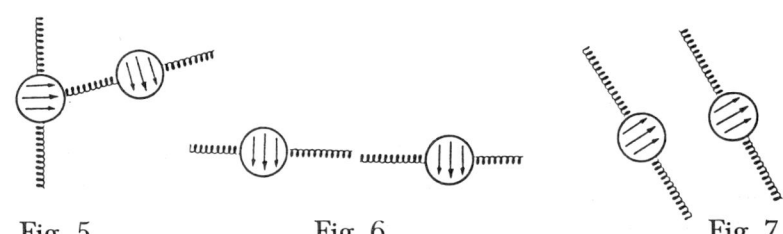

Fig. 5 Fig. 6 Fig. 7

Since the nucleus is spinning, before being pulled, the force of its spin will quickly push aside the gravity line or force the nuclei to move and escape from it so that the gravity line will pass by its side. Another reason is that, due to the spiral nature of the suction line, it automatically bends and directs itself to the side of the nucleus and passes by it. The second possibility is shown in Figure 6 where the suction lines of the two nuclei are directed towards one another head-on. In this case, again the suction lines will quickly push each other aside because they are extremely thin and turning spirally, and the force of one pushes aside the other. The third possibility is that the suction lines of the two nuclei are not directed to each other as shown in Figure 7. In this case, the two nuclei do not attract each other and therefore, do not combine.

Why does the gravitational force act independently of the magnetic force? Moreover, why is the gravitational force is more stable than the magnetic force? The reason the force of gravity is independent from the magnetic force is because their sources are different. The source of gravity is the nucleus, while the source of the magnetic force is the electrons and their samarehs. Heat quickly affects the velocity of electrons around a nucleus, but heat has comparatively little effect on the spin of a nucleus. The stability and independence of the gravitational force is very similar to the stability and independence of gamma rays as compared with the light generated from the motion of electrons and their samarehs. The light from electrons and their samarehs is easily affected by chemical changes, by heating or by motion, whereas gamma rays are not easily affected by these changes. Since gamma rays are from the nucleus itself, and visible light is from electrons and their samarehs, we see why gamma rays are independent from visible light. Similarly, since the gravitational force is from the nucleus itself, the force acts quite independently of the magnetic force.

It was mentioned before that the faster an object spins, the longer and narrower its suction tunnels become. In the case of an object spinning in water, the velocity of the spin is usually less than one hundred revolutions per second, whereas in the case of a nucleus spinning in ether, the velocity of the spin is not one hundred but billions of trillions of revolutions per second. For this reason, the suction tunnels are very long and narrow. Calculations show that with such a fast spin, the suction tunnels could be in the order of hundreds or even thousands of kilometres long. Since the nucleus itself is very small, each suction tunnel becomes an extremely thin line that has a cross section of only a small fraction of the diameter of the nucleus, like a long string pulling and attracting that can be represented by a vector and called a "gravity line". Since in an atom, the space that the nucleus with all its electrons and samarehs combined occupies is very small, through the cross section of an atom then, there can pass a very large number of these gravity lines.

The following will illustrate how the gravitational force of the earth is created.

Fig. 5 shows a circle representing a mid cross-section of the earth with a large number of the nuclei of atoms located inside or on the surface of the earth.

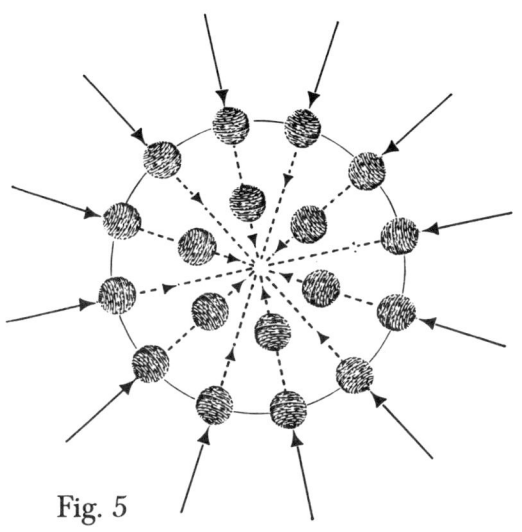

Fig. 5

Notice that the axes of the spins of all the nuclei, all around, pass through the centre of the earth. To see the reason why a spinning nucleus positions itself in such a way that its axis of spin passes through the centre of the earth, let us study why the axis of the spin of a top is always pointed towards the centre of the earth. In my childhood we used to play with a toy called a "top", having a body of a conical or circular shape. Its sharp end was made of metal on which it turned. The top was spun by means of a cord or by a twist of fingers. I noticed that it always spun perpendicular to the ground, and would spin that way until it slowed down, falling eventually to its side. Even if I tried to make it spin at an angle by starting it at an angle, it would automatically change its angle and again spin perpendicular to the ground. The Encyclopaedia Britannica writes about these tops as follows: "If spun with a slant at the start, it will quickly stand upright till halted by friction".[36] This shows that the spin of the top helps set the axis of its spin towards the centre of the earth. Since the nucleus also spins, for the same reasons it

---

[36] - *Encyclopedia Britannica*, 1968, vol. 22, p. 78.

automatically sets its axis of spin towards the centre of the earth.

The arrows that have solid lines, all pointing towards the circle of the earth, show gravity lines where ether is flowing with high speed towards the poles of the nuclei. The dotted lines are gravity lines at the opposite poles where there is no free flow of ether. The force is only pulling the atoms towards the centre of the earth or holding the atoms together. The reason why atoms are pulled towards the centre of the earth is because (looking at the centre) nuclei from all sides are trying to suck ether from all directions. Since there is no free flow of ether from the centre and through a dotted line, the suction activity of a nucleus only helps to pull the nucleus towards the centre of the earth. It is similar to the suction hose of a vacuum cleaner that can suck the small objects and move them towards itself, whereas if there is a very large object in front of it that prevents the free flow of air, then the hose itself is attracted and moves towards the object. As another example, if you are pulling a rope which is fastened to a small object, the object is moved towards you, but if the object is immovable, then by pulling the rope, you yourself are moved towards the object. Hence the lines of gravity, the dotted lines, act as implosive forces that continuously pull atoms towards the centre of the earth. This explains why everything is attracted towards the centre of the earth. For this reason, the direction of the dotted gravity lines are shown to be towards the centre of the earth and not towards the nuclei. The forces of attraction all pointing toward a centre also explain why all the planets and stars are round.

Looking carefully at the circle (Fig. 6), you will notice that outside the circle, the gravitational forces (the arrows) are all directed towards the circle and no gravitational force is directed away from the circle. This further explains why everything is attracted towards the centre of the earth, and why all the planets and stars are round.

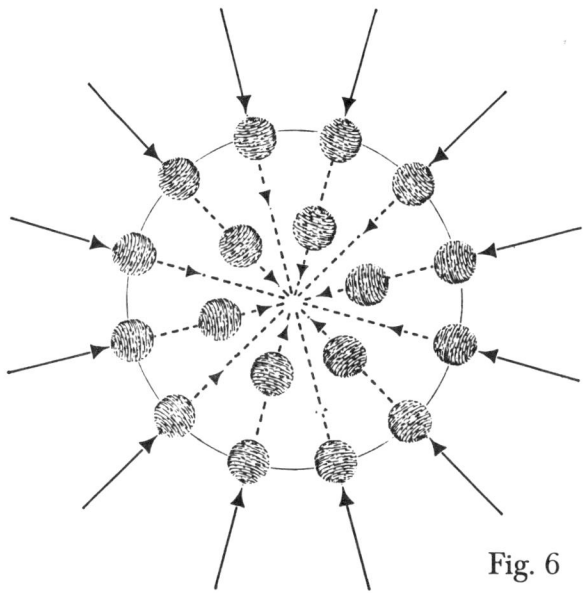

Fig. 6

One might have a question about the gravity of the earth seen in this light by asking the following question: we know that the entire surface of the earth, which is a sphere, exerts a force of attraction on the entire area surrounding the earth. That is why everything on earth has weight. This means that on the entire surface of the earth, all around, there must be some ether continuously flowing into the earth in order to create the force of gravity (weight). From where, then, does the ether that enters the earth come out? The amount of ether going into the earth must be balanced by the amount of ether coming out. In such a case, it would seem as if the ether that is continuously coming out must cancel the force of gravity.

It must be seen, however, that the ether flowing into the earth passes through the narrow tunnels and thus has a much higher speed than ether coming out from the areas about the circumferences of the atoms. The ether that passes through the narrow tunnels, because of its high speed, exerts a much greater influence than ether that moves very slowly. Although the amount of ether going into the earth is equal to the amount of ether coming out, the difference of momentum causes the force of one to be greater than the other. An example will illustrate how the force of ether going into the earth is not cancelled out by the force of the ether coming out. If we put the palm of our hand on a fountain of water which is shooting water upward, we feel the pressure of water on our palm, a pressure that pushes our hand upward (Fig. 7).

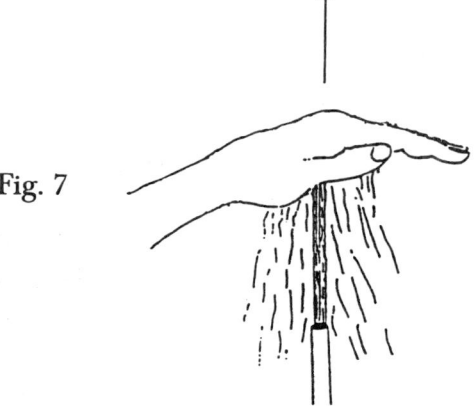

Fig. 7

Although all the water that comes out of the fountain and moves upward after reaching our hand returns towards the ground, yet by its momentum, the water exerts a pressure on our hand. This simple example is the reverse of how the force of gravity is created. The ether that flows into the narrow tunnel exerts a much higher force than the ether coming out of the earth from around the nucleus (Fig. 8).

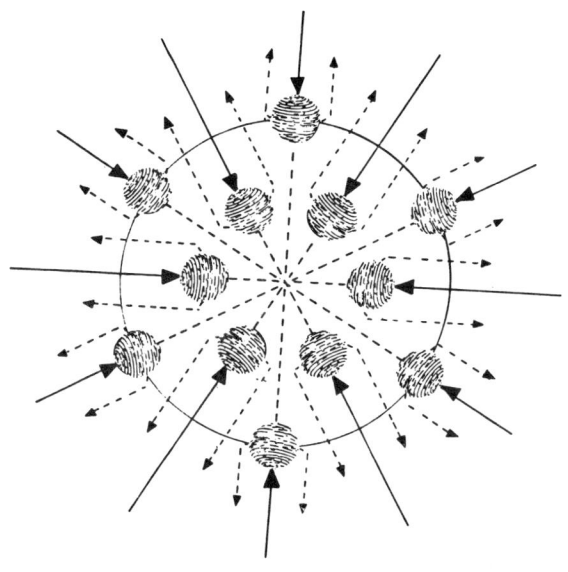

Fig. 8

A further question is this: why do test bodies of different mass, positioned at the same height, fall at the same rate in a vacuum? From the understanding that the force of gravity is from the flow of ether, we then see why different masses in a very strong current of ether are carried with the same speed. Everyone must have seen that in a river with a very strong current, in a given cross section, everything that floats regardless of its size or mass is carried along with the same speed. If the river becomes narrower, everything is then accelerated by the same rate. The same principle applies here. All bodies are carried by the strong current of ether which we call gravity. The gravitational activity of the body in the gravitational field of the earth does not do any work. In other words, the acceleration of a falling body is due to the earth's gravity, and not to that of the body. To see the reason, let us look at Figure 9, which shows only an atom of an object in the gravitational field of the earth.

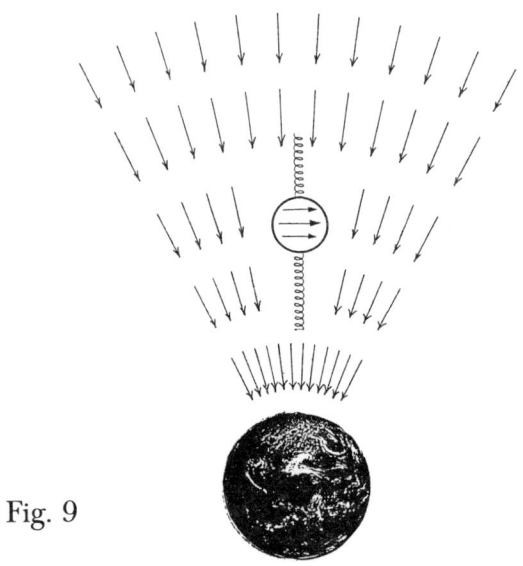

Fig. 9

Notice that all the gravity lines of the earth reaching the atom have only one direction, but the gravity lines of the atom have two opposite directions. For this reason, the gravity line of one pole cancels the opposite pole. Hence the gravitational activity of the body does not do any work. An example will illustrate the case. Imagine a river with a canoe in it. Imagine that two persons sitting opposite to each other, one at each end of the canoe, are paddling exactly in opposite directions. They do no work because the work of one is cancelled by the other and the boat goes only where the current carries it.

It must be borne in mind that a test object falling in the gravitational field will not acquire an instantaneous full speed as that of the current of ether flowing into the earth through the tunnels, because the ether going into the earth is not the only ether in motion. The ether coming out of the earth and the ether that makes up the magnetic field of the earth are examples of ether preventing an object from acquiring an instantaneous full speed.

The reason why an object closer to the ground is subject to a greater gravitational force is because all the lines of gravity pass through the centre of the earth. The closer the object is to the ground, the larger the number of gravity lines passing through it (Fig. 10). The greater the number of gravity lines, the greater the force they have and the greater the acceleration they create.

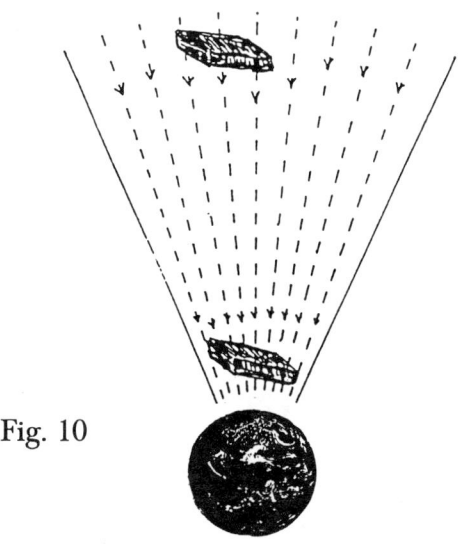

Fig. 10

This also explains Newton's inverse square law of gravity. Newton stated that the gravitational force between two bodies is inversely proportional to the square of the distance between them. This law has been considered to be one of the mysteries in physics. Here the law is no longer a mystery, but a simple logical consequence of the spin of the nucleus in ether.

Consider two spheres r_1 and r_2 around the earth (see Fig. 11).

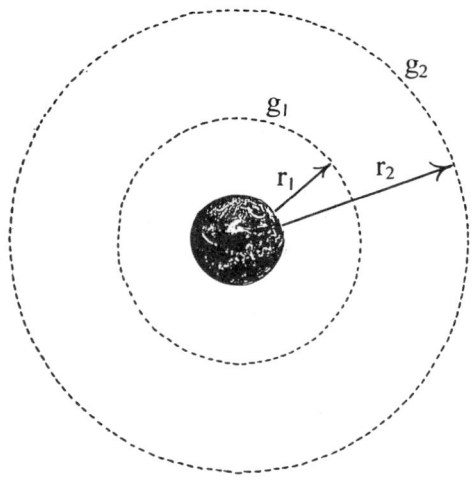

Fig. 11

Gravity lines all pass through the centre of the earth. The number of gravity lines that pass through both spheres are the same. Since the gravity lines all along have uniform cross-sections, then the speed of the current of ether all through a gravity line is the same. This means that the flow of ether per second through the spherical surface $4\pi r_1^2$ must be equal to the flow per second through the surface $4\pi r_2^2$. Since the flow of ether creates the force that we call gravity, hence if g_1 and

g_2 represent the gravitational forces at r_1 and r_2 respectively, then we must have:

$$4 \pi r_1^2 g_1 = 4 \pi r_2^2 g_2 \quad \text{or:} \quad g_1/g_2 = r_2^2/r_1^2$$

This is called the inverse square law of gravity.

Here one can see that ether and the spin of the nucleus provide a simple explanation of gravity that not only answers all questions satisfactorily, but, without any exception, can be demonstrated by simple experiments.

In contrast to the above, one will see that Einstein's explanations about gravity not only cannot be demonstrated by any experiment but contravene many of the known laws of physics.

Today, Einstein's followers have satisfied themselves with Einstein's mathematical explanations of gravity, which, when studied carefully, reveal no answers to any of the fundamental questions such as why everything is attracted towards the centre of the earth. Furthermore, one will notice that the basis of his ideas are so wrong that they make us wonder why Einstein's paper about gravity, known as the General Theory of Relativity, is called a "great scientific work". He does not answer any of the above questions such as: why does gravity exist?- what creates gravitational force? - why is the force of gravity able to act at a distance? and, - why is everything attracted to the centre of the earth? Despite this, his theory has been claimed to be the answer to gravity and has been praised as the greatest ever scientific work. The explanations that we find in Einstein's paper are mixed with a mathematics called "the mathematics of higher-dimensional abstract geometry"[37]. What he has done can be

---

[37] - *McGraw-Hill Encyclopedia of Science and Technology*, (1992), V. 15, p. 288.

characterised as the abuse of mathematics, a jungle of confusing ideas that admittedly, no one can translate into real geometry or anything related to our physical world, as they remain only in an abstract form in the "fourth dimension" that even the imagination cannot fathom.

In order for the reader to be in a better position to judge, the following is a summary of what a very reputable scientific encyclopaedia has described as Einstein's explanation of gravity. This summary is then followed by another written by Einstein himself. These enable the reader to compare and judge for him or herself whether or not Einstein has given any answer to anything related to gravity, and to see whose ideas make more sense, Einstein's or the explanations presented in this book, all of which can be demonstrated by simple experiments.

General relativity is the geometric theory of gravitation developed by Einstein in 1915. . . . Consider an observer in the gravitational field of the earth and another in an accelerating elevator or rocket in free space (Fig. 2). If both drop a test body, they will observe it to accelerate relative to the floor....Einstein was greatly impressed by this fact and elevated it to a general principle, the principle of equivalence: the principle states that on a local scale all physical effects of a gravitational field are indistinguishable from the physical effect of an accelerated coordinate system. . . . . .The close connection between gravity and the accelerating coordinate system convinced Einstein that gravity is fundamentally a geometric phenomenon. Because of this, it is naturally described by the mathematics of higher-dimensional abstract geometry. This geometry involves systems of equations called tensor equations, that are manifestly independent of coordinate systems. Only tensors in the four dimensional space-time of relativity need to be considered. Suppose that the points in four-dimensional space-time are labelled by two essentially arbitrary systems of coordinates...[38]

Then follow some other very unreasonable assumptions and complicated mathematical manipulations, series of approximations, hundreds of equations, terminologies, formulas and their interpretations. The mathematics only hide the problems and serve as a jungle with pitfalls in which anyone who falls in most probably will become lost and disoriented in such a way that he completely loses his perspective. In order not to lose the reader's overall perspective, the mathematical aspects have not been printed here.

Einstein himself summarises his theory as follows:

WE SUMMARIZE:
.... The theory attacks the problem of gravitation and formulates new structure laws for the gravitational field. It forces us to analyze the role played by geometry in the description of the physical world. It regards the fact that gravitational and inertial mass are equal, as essential and not merely accidental, as in classical mechanics. The experimental consequences of the general relativity theory differs only slightly from those of classical mechanics. They stand the test of experiment well wherever comparison is possible. But the strength of

---

[38] *McGraw-Hill Encyclopedia of Science & Technology*, (1992), V.15, p. 287.

the theory lies in its inner consistency and the simplicity of its fundamental assumptions.[39]

Stephen Hawking, a supporter of Einstein, summarises Einstein's theory and writes:

Einstein made a number of unsuccessful attempts between 1908 and 1914 to find a theory of gravity that was consistent with special relativity. Finally, in 1915, he proposed what we now call the general theory of relativity.

Einstein made the revolutionary suggestion that gravity is not a force like others, but is a consequence of the fact that space-time is not flat, as had been previously assumed: it is curved, or "warped," by the distribution of mass and energy in it. Bodies like the earth are not made to move on curved orbits by a force called gravity; instead they follow the nearest thing to a straight path in a curved space, which is called geodesic.[40]

In brief, according to Einstein, gravity exists because space-time bends or warps. How could anyone accept the idea that objects have weight because time bends and space warps?

To see how irrational is this idea, imagine you are holding two cubes, exactly the same size, one made of lead and the other made of aluminum. According to Einstein, the two cubes have different weights because of the curvature of time

---

[39] - . Albert Einstein and Leopold Infeld, *The Evolution of Physics*, Simon and Schuster, New York, (1938), p.260.

[40] - Stephen W. Hawking, *A Brief History of Time*, Bantam Books, Toronto, New York,(1988), p. 29.

and space. Bearing in mind that as you are holding the cubes the time for both cubes is exactly the same, and that since the cubes have the same size, the amount of space for both cubes is also exactly the same. How can one accept that the cubes have different weights because time and space curve? How can this illogic be accepted as science?

On the other hand, the simple understanding that the nuclei of atoms spin is very logical and natural. Science has already proven that electrons revolve around the nucleus. Science has also proven that electrons indeed do spin. Is it then an unreasonable proposition that the nucleus also spins? If one carefully investigates the evidence provided, it will be found to be overwhelmingly clear and irrefutable.

It is also known and can easily be demonstrated that the spin of a spherical object in water or air will create a force with identical characteristics to the force created by the spin of the nucleus in ether. In contrast to this, one can see that Einstein's explanations of gravity, that is to say, the curvature of time and space, and four-dimensional geometry cannot be experimentally proven and do not even agree with common sense.

The following are some examples of the misrepresentations which are contained in Einstein's General Theory of Relativity.

1- If we study the so-called "principle of equivalence", we see that the whole principle is erroneous because the physical effect caused by gravity is not the same as that caused by an accelerating body. Acceleration created by gravity increases as the body gets closer to the earth, even in a local scale, whereas the physical effect caused by an accelerating body does not increase. How can we call these two essentially different phenomena equivalent? On what basis was this idea elevated to a principle?

2- If 100 pounds of force from one source is equivalent to that of another source, this gives us no answer as to how the forces were created. What does equivalence have to do with the nature of the gravitational force and how gravity acts at a distance ?

3 - Assuming the two phenomena to be equivalent, how can we conclude that gravity is fundamentally a geometric phenomenon? What does gravity have to do with geometry? Can geometry create a force? What kind of geometry (with four dimensions) is it that no mind can comprehend?

4- How can a system of mathematics, all of whose parameters are based on arbitrary assumptions, be termed a proof of the true nature of gravity?

5- How can mathematics, which is only a means to finding an answer, be called the answer?

6- The idea of the fourth dimension has had no scientific foundation,[41] has never been proven and therefore cannot be called science. Rather, it belongs to the realm of fiction.

7- It is assumed that the nature of time is exactly the same as that of space and that time turns into space or space becomes time. If you seriously ponder these possibilities, they dissolve into vain imagination. It is also assumed that four-dimensional coordinates can exist by interchanging the time coordinate with those of space.

---

[41] - See the section called "Einstein's Claim Against Ether".

8- All the ideas are based upon hypotheses that are built upon hypotheses.

9- All the explanations lack the basic and important answers about gravity.

Finally, the following is how the Dictionary of Scientific Biography summarises Einstein's explanation of gravity:
> *Gravity is a universal manifestation because it is the property of time-space, and hence everything that is in space-time (which is literally everything) must experience it.*[42]

Does this very superficial explanation solve anything about gravity?

Some scientists believe that the acceptance of Einstein's paper and ideas came about by mathematicians and not by physicists, because physicists reading Einstein's paper would not understand anything.[43] Furthermore, the name that Einstein had given to his paper, the "General Theory of Relativity", has created in the minds of many the false idea that the theory of relativity belongs to Einstein, when in fact the actual theory of relativity belongs to Lorenz and not Einstein.[44]

Herbert Dingle, a distinguished professor of the History and Philosophy of Science at the University of London, writes:

---

[42] - A. T. Macrobius & K. F. Naumann, *Dictionary of Scientific Biography*, Volume. IV, p. 331.

[43] - Stephen W. Hawking, *A Brief History of Time*, Bantam Books, Toronto, New York,(1988), p. 83.

[44] For details, see Chapter 31, "Einstein's Rejection of Ether".

> . . . with the apparent success in 1919 of Einstein's general theory, with its then quite new and terrifying mathematical machinery of tensor calculus, came the fatal climax. Almost overnight 'the relativity theory of Lorenz' became 'Einstein's special relativity theory', and it was immediately hailed as such by the mathematical experts. The established physicists, therefore, had to face the alternatives of accepting, without understanding, the metaphysics of the newly christened 'Einstein's special theory', or mastering tensor calculus sufficiently to show that the so-called general relativity theory was not necessarily a generalisation of the earlier Einstein form of the 'relativity theory', and therefore carried with it no justification of 'Einstein's special relativity' (this is explained in detail in Part Two). They chose the former alternative. They gave up trying to understand the whole business, surrendered the use of their intelligence, and accepted passively whatever apparent absurdities the mathematicians put before them.[45]

After Einstein's paper, physics was changed into pure mathematics. Rudakov writes:

> In the last sixty years physics has been enslaved by theoreticians who have succeeded in abolishing physical reality and replacing it with an empty and barren mathematical formalism. The new physicist no longer studies nature and describes what he has observed in physically meaningful terms. He sits at

---

[45] - Herbert Dingle, *Science at the Crossroads*, Martin Brian & O'Keefe, London, (1972), p. 95.

the desk, manipulates abstract symbols and figures, and communicates what the universe is like and how it ought to behave in the form of equations which are comprehensible only to a small and exclusive group of theoreticians like himself.[46]

It is very clear that by the abuse of mathematics, a few[47] were able to mislead mankind and monopolise the understanding of the science. Complicated mathematics was used as a tool, making people believe that what Einstein had suggested was true. Herbert Dingle, on the modern misapplication of mathematics writes:

A language is a medium for expressing ideas, and it is just as capable of expressing false ideas as true ones. The fact, therefore, that something can be expressed with rigorous mathematical exactitude tells us nothing at all about its truth, i.e. about its relation to nature, or to what we can experience.

The most dangerous intellectual error of modern science, with which this book is concerned, lies in the fact that this has been overlooked. Mathematics is an immensely more powerful tool than the Aristotelian syllogism, and its use as a language in which to express the facts of experience has been so successful that the idea has crept unperceived into the minds of physicists that whatever it says must be true. . . . The fact is, however, that mathematical truths are far more general than physical truths: that is to say, the symbols that

---

[46] - N. Rudakov, *Fiction Stranger than Truth*, Australia, published by N. Rudakov, (1981), p. 1.

[47] - Stephen W. Hawking, *A Brief History of Time*, Bantam Books, Toronto, New York,(1988), p. 174.

compose a mathematical expression may, with equal mathematical correctness, correspond both to that which is observable and that which is purely imaginary or even unimaginable. If therefore, we start with a mathematical expression, and infer that there must be something in nature corresponding to it, we do in principle just what the pre-scientific philosophers did when they assumed that nature must obey their axioms, but its immensely greater power for both good and evil makes the consequences of its misapplication immensely more serious. [48]

    The eminent French mathematician, Painleve, in his investigation of Einstein's relativity papers, found many problems with Einstein's interpretations of the relativity formulas. Painleve concludes that it is pure imagination to pretend to draw conclusions such as Einstein does.[49]

    There have been numerous articles written by scientists criticizing Einstein's works. However, since none of them could come up with an alternative solution to the gravity problem, they were not successful in refuting Einstein's ideas. Here we see that the concept of ether not only solves the problem of gravity, but also provides the answer for the magnetic and electric forces. It must be borne in mind that Einstein, by abandoning ether, not only did not solve the gravity problem, but actually turned many solutions into problems. Up until Einstein's time, the electric and magnetic forces were known to be from the flow of ether.

---

[48] - Herbert Dingle, *Science at the Crossroads*, Martin Brian & O'Keefe, London, (1972), p. 30.

[49] - P. Painleve, "Classical Mechanics and the Theory of Relativity", *Science Abstracts: Section A-Physics*. Vol. 25, Part 3. (March 31, 1922), p. 170.

The only problem that remained was the problem of gravity. Following Einstein, none of these forces could be explained. Gravity, electricity and magnetism sank back into the realm of the inexplicable.

Chapter 13

Inertia

The existence of ether also explains inertia. Inertia is a force created by the resistance of ether to acceleration or deceleration. An example will illustrate how inertia is created by acceleration.

If you have tried running fast in water, you will have noticed that water resists the sudden acceleration or deceleration of your feet. Similarly, ether resists acceleration and deceleration. This resistive force is called 'inertia'. In the case of water, the sensation of the resistive force of water is felt on the surface of the skin. However, the resistive force of ether (inertia) on our bodies is not felt on the skin, but rather over the entire body. This is because, in the case of water, the force of water is mainly on the surface of the skin, as water molecules cannot penetrate the skin's surface. In contrast, ether particles easily pass through different layers of skin and through our body. This is why the force of inertia is felt on the entire body.

A body in acceleration not only creates disturbances in front of and within the moving body, but also creates disturbances all around the body and at a distance to the body. Since while an object is accelerating, the ether passes through the object, the flow of ether through the body may cause the object to become magnetised. This is exactly the reason why an unmagnetised object made of steel, if accelerated, very quickly becomes magnetised. You must have noticed, if you have used a high-speed drill, that after its use, a drill-bit made of steel becomes a very strong magnet. It must be noted that while a drill-bit is turning it is in a state of acceleration. The flow of ether through the body is in reality the same thing as the passing of a magnetic field through the body. This is because wherever ether flows there is a magnetic field, and a steel bar near a magnet becomes magnetised. Experiments have also shown that an iron bar in acceleration will also act as a magnet.

Chapter 14

The Electric Field

Here it will be illustrated why an electron creates a field around itself, why it is attracted to the nucleus and why it repels another electron. It will also be shown what a negative charge and a positive charge are, and finally, why electrical current flows from higher to lower potential.

In 1890, A. Schuster, and later in 1897, J. J. Thomson, were able to measure the charge of an electron and also to prove that electrons have mass. The mass was found to be about 1/1837 that of a hydrogen atom[50]. In 1925, G. E Uhlenbeck and S. A. Goudsmit, by studying extensively the atomic spectra, came to the conclusion that electrons, like tops, spin, and as a result have angular momentum[51]. Furthermore, they concluded that electrons also have magnetic momentum which is responsible for the magnetic behaviour of materials such as iron.

In Chapter 8, "The Samareh (Electromoon)", we studied how the motion of the samareh is responsible for the magnetic behaviour of materials. In this chapter, we will study how the spin of electrons in ether is responsible for the charge of an electron.

In Chapter 11, "The Nucleus and Its Spin", it was necessary to prove that the nucleus of an atom spins, and that the spin in ether is responsible for the force of gravity. Here, fortunately, there is no need to prove that the electron spins, because the spin of electrons has already been proven. However, it must be proven that the spin of the electron in ether is responsible for the charge of the electron, and that

---

[50] - See page 45.

[51] - *Encyclopedia Britannica*, 1967, Vol. 8, p. 243-244.

because of the spin, electrons repel one another. In Chapter 12, "Gravity", it was also shown that due to the spin in ether, the axis of the spin of the nucleus is always set towards the centre of the earth. Since electrons also spin, the axis of their spin is also set towards the centre of the earth.

After a very careful study of electricity and the behaviour of an electron around an atom, it becomes clear that there is perfect symmetry between an atom and the solar system, the electron resembling the planets that revolve around the sun. Similar to the way that the atmosphere of the earth spins with the earth as one body, together having one axis of spin, the ether in the immediate vicinity of the electron also spins with the electron as if they make one body.

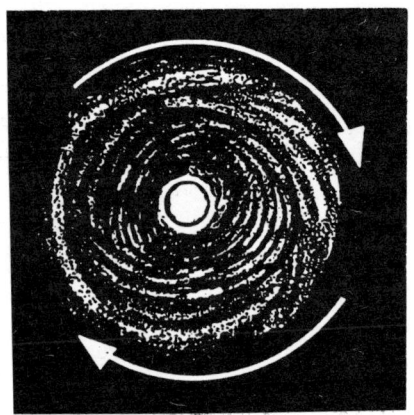

Fig. 1

Fig. 1 shows an electron and the ether about it. The electron and its atmosphere of ether together spin about one axis. It must be borne in mind that the ether that spins with an electron is not the same as the ether that circulates as a result of the samareh's spiral motion. The samareh is relatively far from the electron and its surrounding ether as the moon is relatively far from the earth's atmosphere. Therefore, the spiral motion of ether due to the samareh's motion has nothing to do with the ether that shares the electron's spin. Since the direction of the spiral motion of the samareh is different from that of the spin of the electron, the force arising from the samareh's motion does not interfere with the force arising from the electron's spin.

Fig. 2 shows two electrons approaching each other.

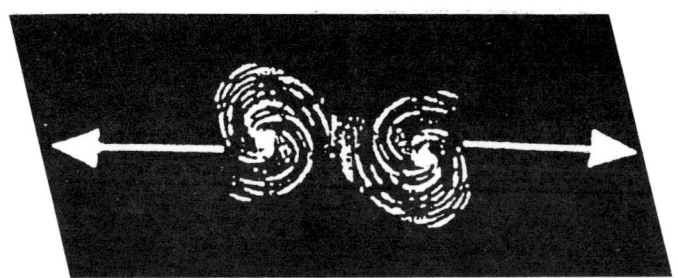

Fig. 2

Although the direction of the circulating ether of each electron is clockwise, yet at the point of contact it is in opposite directions. Hence they create friction and pressure, a pressure that causes the electrons to repel each other. The pressure between free electrons is responsible for the electrons moving from a very crowded area to a less crowded area. The area where there are high numbers of electrons is called "high potential", and the area where there are less electrons is called "lower potential". Of course, the area in which one electron can apply a force to another is called the "electric field".

Fig. 3 shows an electron in the vicinity of the nucleus of an element.

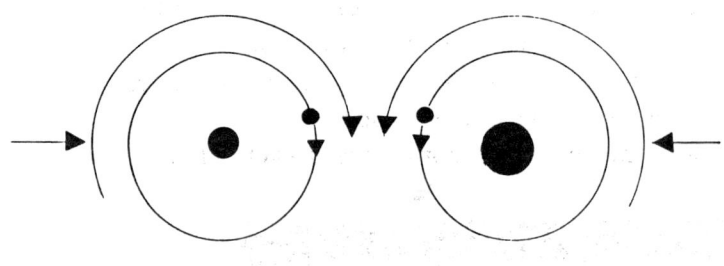

Fig. 3

Since both the nucleus and the electron spin, and the nucleus spins in the opposite direction to that of the electron, the revolving ether of both the electron and the nucleus at the point of contact move in the same direction, thereby reinforcing and attracting each other. The phenomenon can also be demonstrated by using two tops, one large and one small, both spinning near each other but in different directions. Fig. 4 shows two tops spinning in different directions. The circulating wind generated by the spin of the two tops creates a force of attraction between the two. Since one of the tops is bigger than the other, the smaller one will start revolving around the larger one. This explains why electrons revolve around the nucleus and the planets revolve around the sun.

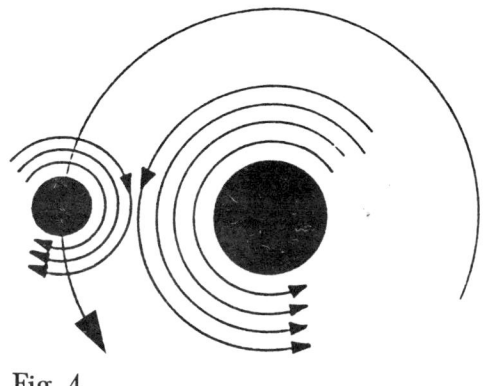

Fig. 4

Here we see a very simple and logical explanation that has absolutely no problems. In contrast, when one studies what the scientists of today say about the electric field and why electrons repel one another, one finds that there is no explanation; there is no theory behind it at all. These important questions have been avoided altogether. This is because Einstein removed the essential ingredient, ether, without which they could not come up with anything that could explain even superficially why these forces are able to attract or repel from a distance.

Chapter 15

Radiation as Waves of Ether

Before Einstein's mystification of science, physicists understood that all forms of radiation such as radio waves and light waves are, in reality, the wave and vibration of ether. Einstein changed everything by claiming that these forms of radiation are "electromagnetic waves". At one moment they are electric fields, at another, magnetic fields, which for unknown reasons are able to interchange from one form to another and that somehow, for unknown reasons, can spread in empty space without the medium of ether. Einstein also claimed that light behaves as both a wave and a particle. Later on, it will be illustrated that Einstein, by misrepresentation and manipulation of mathematics, was able to mislead scientists into accepting his contradictory wave and particle theories of light.

When one looks at the spectrum of all radiation emanating from an atom, namely heat, light, x-rays, and gamma rays, one can see a series of distinct radiations with distinct characteristics that come from distinct parts of the atom. Heat, having the longest wavelength, up to gamma waves which have the shortest wavelength, are the two extremes. Here it will be shown that vibrations of the subatomic particles in ether are responsible for each.

Heat is caused by the vibration of atoms and molecules. That is why friction between two surfaces generates heat. When an atom is heated, the electrons move faster around the nucleus. The fast orbit of the electron causes the atom to vibrate. The vibration of the atom in ether generates waves which we call 'heat waves'. Microwaves are, in reality, heat waves, but with exact frequencies which immediately excite an atom into vibration and resonance. Light waves with higher frequencies are created by the vibration of the samareh. When a samareh is in an excited state, it generates light. A samareh moves around both the electron and the nucleus. The fast orbit of a samareh around an electron generates light. The fast orbit of an electron causes the vibration of the atom and generates heat. That is why the fast motion of an electron and samareh not only generate light, but also heat, as the whole atom is in a state of excitation and vibration. The impact of a fast-moving electron with a hard surface generates x-rays. This phenomenon is very similar to a sound wave generated by a ball striking a hard surface. Gamma rays, which are generated by the spin of the nucleus itself, have the shortest frequencies (Fig. 1). A very fast spin of the nucleus is accompanied with vibration. The vibration of a nucleus in ether in turn generates gamma rays.

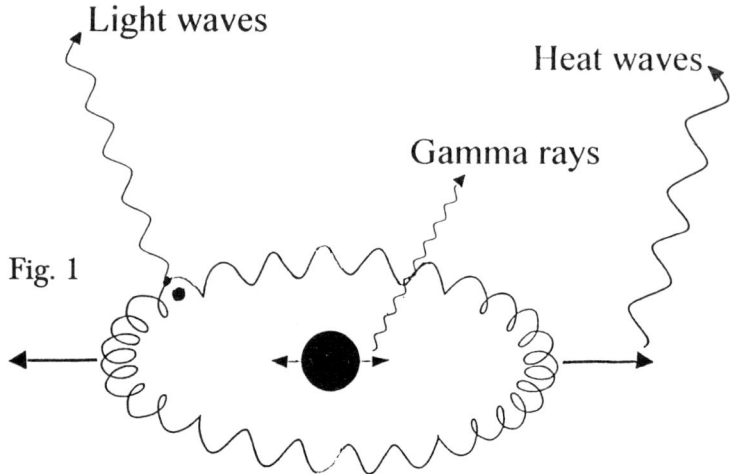

Fig. 1

The discovery of the samareh and the spin of the nucleus are the missing pieces of the puzzle that connect the series of different forms of radiation to their causes.

There are other sources of radiation which must be noted here. Radio waves, the light of stars, comets, interstellar dust activity, the fast motion of ether itself in space, the gigantic invisible magnetic waves generated by the activity, the vibration or the spin of stars or by the motion of celestial bodies such as luminous and non-luminous spheres and comets, are all waves of ether. Some of these waves have very long wavelengths with frequencies as low as 1 cycle per decade, making them very hard to detect. Radio waves which are also waves of ether are caused by disturbances generated by changing magnetic or electric fields in ether. These disturbances as waves are simply propagated in space by the medium of ether. The following will illustrate how electrical disturbances in ether generate radio waves.

It was shown that the motion of an electron (with its samareh) from one point to another causes the surrounding ether to circulate in a right-hand screw direction. In other words, the current in a conductor creates a magnetic field in a right hand screw direction (Fig. 2).

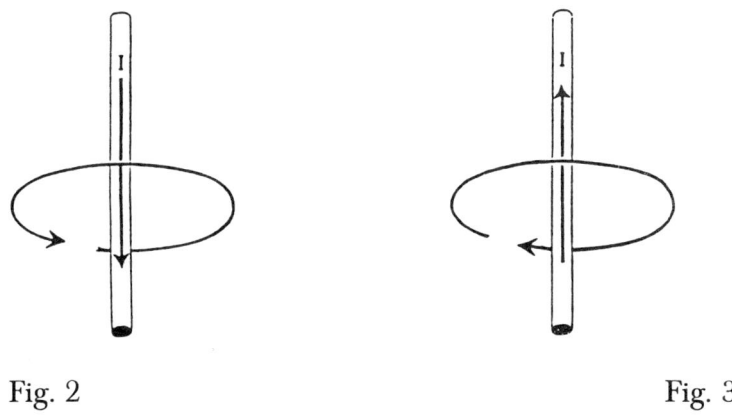

Fig. 2   Fig. 3

In Fig. 3, the direction of the current is reversed, and the direction of the circulation of ether is also reversed.

In an alternating current in which the direction of the current is constantly changing, the direction of ether circulating around the conductor is also constantly changing. This constant change in the direction of ether from clockwise to counter clockwise and vice versa becomes a disturbance or wave in ether which is then transmitted by the medium of ether in space as a radio wave (Fig. 4).

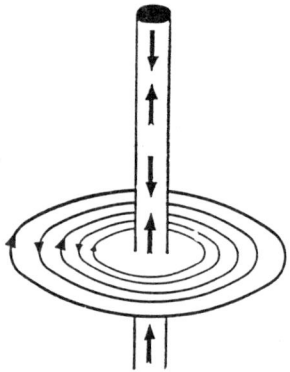

Fig. 4

The frequency of the radio wave depends on how fast the direction of circulation of ether is changed back and forth in a given time period. An example will illustrate how a wave is created and propagated through space. Suppose we have a bar with a plate attached to it (Fig. 5).

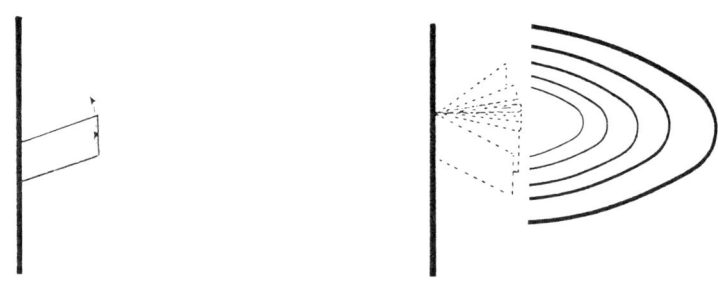

Fig. 5   Fig. 6

If the bar turns clockwise and counterclockwise with a very high frequency (Fig. 6), the plate that is attached to the bar moves back and forth (vibrates) and creates waves in air. The wave is then transmitted by the medium of air in space. If the number of vibrations per second is within the audible frequency, our ears will hear them as sound waves. Similarly, the circulation of ether back and forth creates waves in ether that we call radio waves.

DIFFERENT RADIATIONS

Fig. 7 is a chart of the wave spectrum. All are basically waves of ether. Although waves with different frequencies are created by different methods, all are waves of the same medium.

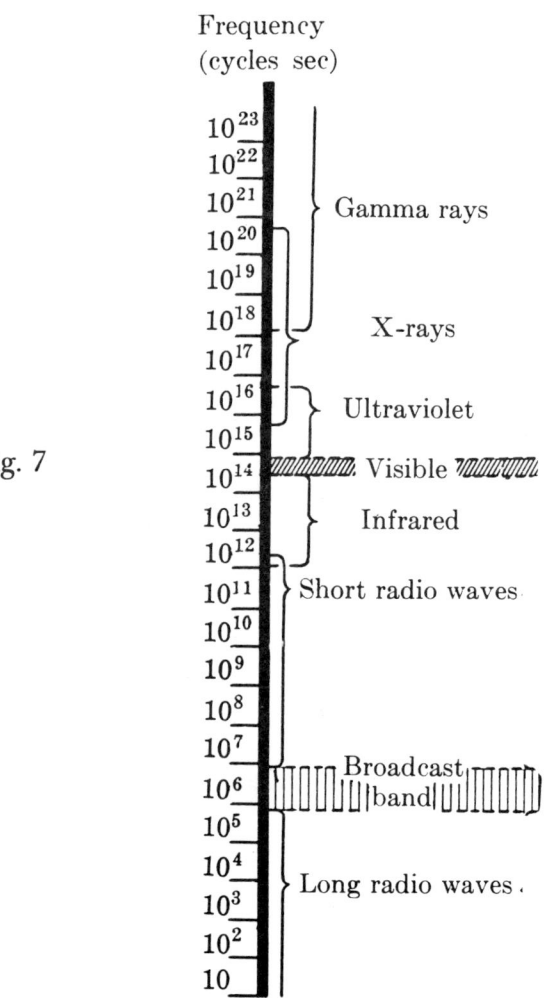

Fig. 7

Chapter 16

Unitary Field

Einstein spent much of his life trying to invent a unified field theory that would explain the three forces of nature, namely gravity, electric and magnetic forces, by one generalised geometrical structure based on his theory of relativity. His efforts up to the last days of his life dissipated in vain. Is this not a good reason to question whether the basis of his theories may be wrong?

Motions of the nucleus, electron and samareh and their interactions with ether cause these phenomena. They also provide a very simple explanation of all forms of radiation.

In previous chapters, we proposed that the spin of the nucleus generates gamma rays, while the spiral motion of the samareh generates light, and that the fast orbital motion of electrons causes an atom to vibrate and generate heat. (Fig. 1).

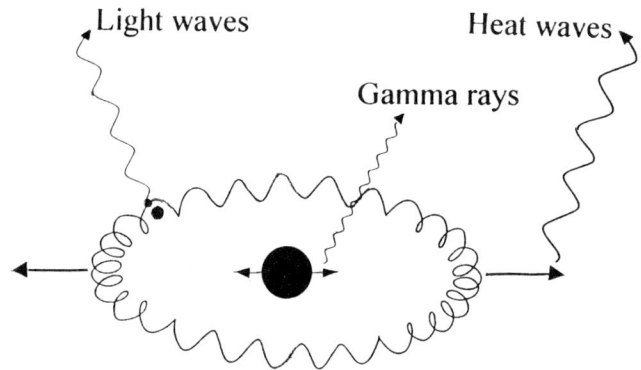

Fig. 1

Similarly, the spin of the nucleus creates gravity, while the spin of the electron creates electric forces. Moreover, the spiral motion of the samareh around the nucleus generates the magnetic force and makes of an atom a magnet (Fig. 2). This explains why the electric field and magnetic forces are in reality the same thing.

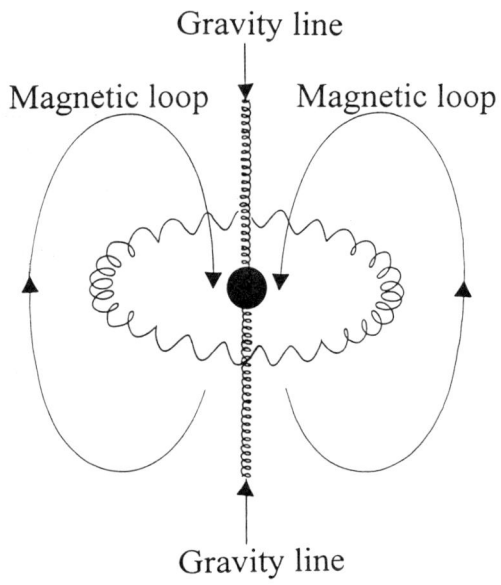

Fig. 2

Although the different forms of radiation are basically the same in that they are all waves of ether, yet they have distinctive characteristics because they are generated by distinct processes. Similarly, the gravitational, electric and magnetic forces are all basically the same. They are only forces arising from the flow of ether, but because they are created by different sources in different ways, the flow of ether in each case has different characteristics, and hence different effects.

Chapter 17

The Transmission of Electrical Energy in a Conductor

In previous chapters, we saw how light and heat are created from the wave and vibration of ether. In this chapter, we will examine how electricity is generated from the wave and vibration of ether.

The wave and vibration of ether moving along a conductor causes some of the electrons to move along with it. The movement of an electron along a conductor is called 'electrical current'.

Within a conductor, the distances between adjacent nuclei are relatively long. The total size that each nucleus with all its electrons and their moons combined will have is only a fraction of the space that the whole atom occupies. The nucleus and its electrons are not located in empty space. The so-called empty space within atoms is filled with ether.

James Clerk Maxwell, the well-known scientist of the nineteenth century, and a supporter of Faraday's theory, wrote in his famous paper on the electromagnetic theory of light:

> *It appears therefore that certain phenomena in electricity and magnetism lead to the same conclusion as those of optics, namely, that there is an aetherial medium pervading all bodies* . . .[52]

The ether that pervades a conductor is the medium which transmits disturbances, or waves, along the conductor. Bear in mind that ether is also the medium which transmits light waves. For these reasons, the electrical disturbances,

---

[52] - J. Clerk Maxwell, "A Dynamical Theory of the Electromagnetic Field", *Phil. Mag.* December 8, 1864, p. 464. For a copy see Appendix.

'signals', in a conductor, travel with the speed of light. The waves of ether that move along the wire cause the electrons also to move along. However, the electrons do not move with the speed of light. The motion of electrons along a conductor is called 'drift velocity'. The drift velocity of an electron is only a small fraction of the velocity of light.

The phenomenon of the wave of ether along a conductor causing some of the electrons to break from their atoms and move along is very similar to a phenomenon called the "photoelectric effect". In the phenomenon of the photoelectric effect, light waves, which are in reality ether waves, cause some electrons to break from their atoms and move away from the metal surface. In both phenomena, ether waves are responsible for the motion of the electrons. Since an electrical disturbance within a conductor travels with the speed of light, this in itself is an indication of the existence of ether, because the disturbance must be carried by a medium. Otherwise, there is no possibility that the speed of a signal should be equal to the velocity of light.

If we assume that a series of collisions between electrons is responsible for the transmission of electrical signals in the conductor, then as the electrons do not touch each other and are very much separated, there would be a relatively long time lag between each collision. The addition of many such time lags would make the signal velocity so slow that it would be a small fraction of the velocity of light. The fact that the signals move with the velocity of light shows that the signals must be transmitted by the medium of ether. This is exactly the triumph which Maxwell was able to prove.[53]

---

[53] - Ibid.

Chapter 18

The Motion of Electrons in Ether

Einstein's orthodox followers assume that when an electron jumps from one wire to another, the electron, somehow and for no reason, shoots off countless photons in all directions. Some of these photons reach our eyes and are seen as a light or spark.

Here it will be reasoned that the fast motion of an electron in ether generates light.

When the velocity of an electron in ether is sufficiently high, the electron creates a large amount of friction and vibration in ether. Therefore, the light of a spark is caused, not by the electron itself, but rather by the friction created from its fast motion in ether. When a spark jumps between two wires, there is a rapid motion of electrons from one wire to another. The electrons on their way create friction, excitation and vibration in ether, which we see as a light or spark. Again, it is not the electrons themselves that give light - no "photons" leave the electrons - but rather their motion in ether and the friction they produce that becomes light. If we assume that the light of a spark is caused by photons leaving electrons, then during the short distance that an electron moves from one electrode to another, it must give up countless trillions of photons in all directions, which could not possibly be true.

If the space between the two electrodes is filled with air, the spark is accompanied by sound. This occurs because the electrons ionise the air as they move through it. The ionised air is then attracted to the opposite electrode. As a result, there is a sudden surge of ionised atoms from one wire to another. This sudden disturbance creates sound waves. The rapid motion of ions from one wire to another also creates light and enhances the light of the spark. That is why the light of a spark is enhanced by the presence of air. Electrons moving through a strong magnetic field also create light. This phenomenon has been observed in particle accelerators called Synchrotrons. Since the magnetic field can be easily understood if we call it the etheric field, or where ether is flowing, then we can easily see how the electron moving through a strong current of ether creates friction and vibration and generates a series of radiations, including light and x-rays.

Here one can see that the motion of a high-speed electron in ether is similar to that of a comet. The electron's motion in ether creates a bright coma[54] and a tail similar to that of a comet. The subject of comets will be discussed in greater detail in the next chapter.

---

[54] - The coma is the bright envelope surrounding the nucleus of a comet.

Chapter 19

Comets and Ether

From inside an airplane looking down over a lake, everyone has seen the tail waves, or "wakes", of motor boats on water. The white tail that we see is created by friction, excitation and vibrations in water. The motion of a comet in ether is very similar to the motion of a boat in water. The tail of a comet is created by the friction, excitation and vibration of ether.

An object moving in water not only creates a tail wave, but also creates shock waves at a distance ahead of it. The object also drags along some of the water with it. The shock wave which is created at a distance ahead of the moving object is called the "bow shock". This can be clearly demonstrated, especially if the object is only half submerged in water (Fig 1).

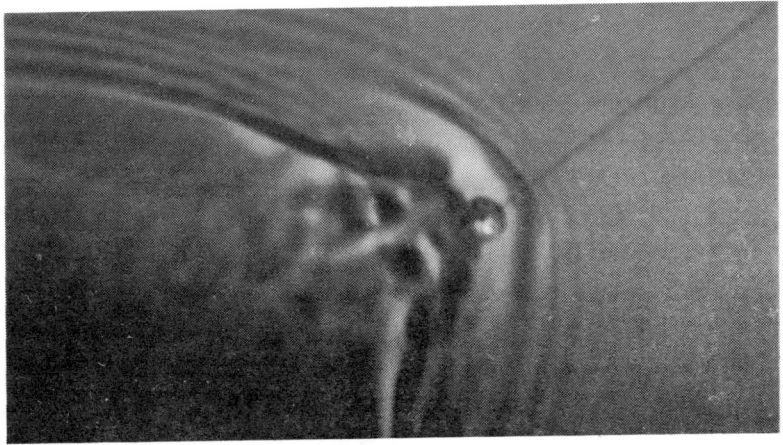

Fig. 1

Looking at the overall waves created, one can see that the pattern has the configuration of a comet.

Fig. 2 shows a sketch of a comet, such as the Great Comet of 1843. Fig. 3 is a sketch of the waves created by the motion of an object in water.

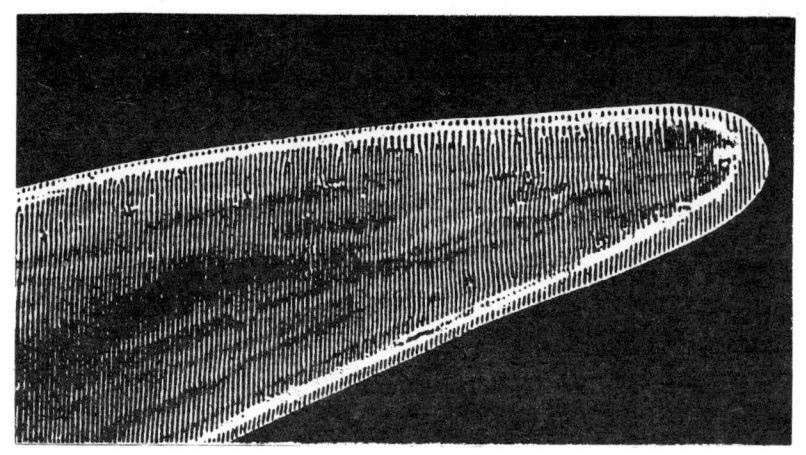

Fig. 2

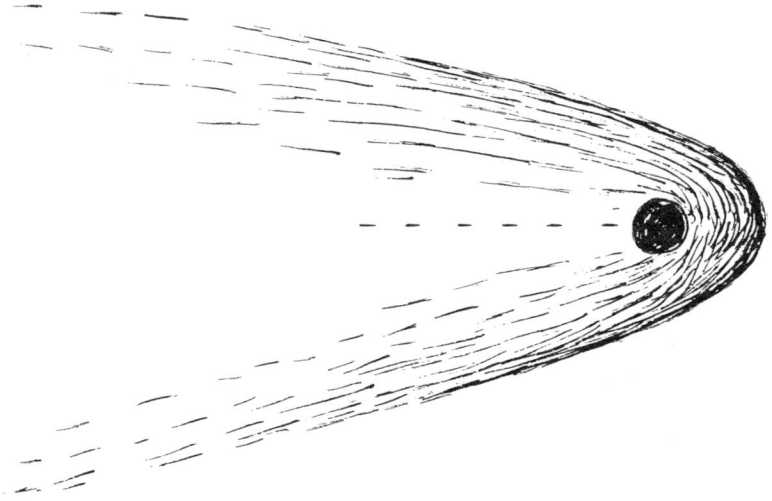

Fig. 3

Notice the similarities between these two sketches. It will be shown in later chapters that the motion of the earth in ether creates similar waves, namely the bow shock and tail waves, but because the earth moves relatively slowly, the waves are not luminous.

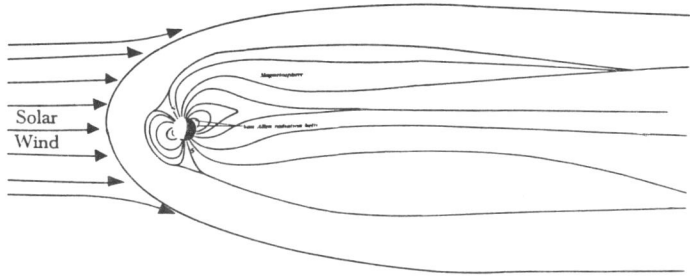

When the velocity of an object in water is slow, friction between different layers of moving and stationary water is not sufficient to create noticeable vibrations in the water. However, when the velocity is very high, friction creates intense vibrations in the water which we see as a white-coloured tail and hear as sound waves. Similarly, comets drag some of the ether along their path and create a coma, a bow shock and tail waves. Since the velocity of a comet relative to ether is high, the friction between the moving and stationary ether creates vibrations in ether. Some of these vibrations are in the visual frequency range and are seen as the light of the comet. The friction and vibrations are greatly enhanced when the comet encounters the solar wind.

The length of the tail of a comet depends on its speed: the higher the speed, the longer its tail. You must have noticed that the faster a boat moves, the longer and more excited the wake becomes. Similarly, the faster a comet moves, the longer and brighter is its tail.

Here, we will review in brief some theories about comets and the problems associated with them. Then it will become clear to the reader that the problems prove these theories wrong.

Some theories explain comets on the following two assumptions:

A - The comet's tail forms from the disintegration of its nucleus. The tail is composed of ions, smoke, dust, ice particles, or even bits and pieces of the comet. The light from the tail of a comet is just a reflection of the light of the sun.

B - Space is without ether.

When one examines each of the basic assumptions, one finds there are many problems and questions that cannot be answered. For example:

1 - Looking at a comet, one observes that the width of its tail uniformly expands as the distance increases from the nucleus. This is the exact configuration of the tail waves, or "wakes", of ships moving in water. If however we assume, as scientists maintain, that the tail arises from particles leaving the comet, the tail should be narrow and long, as in figure 4.

Fig. 4

2 - If we assume that the light of a comet's tail is from the disintegration of its nucleus, then there is no reason why Halley's comet should have a large head. It should have no coma, because the size of its nucleus is very small, only about 15 kilometres long. For a small mass with a small gravity, there is no reason why gas or ice particles etc., should accompany the nucleus at such very large distances. The shape of the comet should be pointed, with no coma or bow shock.

3 - Moreover, if a comet's tail forms from the disintegration of its nucleus, then there should be no bow shocks,[55] especially at very large distances ahead of the nucleus. The Soviet Union's spacecraft Vega 1 recorded the bow shock of Halley's comet at 400,000 kilometers in advance of the nucleus. What could be the reason for a small mass to create a shock wave 400,000 kilometers away from the comet? This can only be explained by ether. Experiments have shown that a moving object in water does create bow shocks a relatively large distance from the comet. Similarly, the small nucleus of a comet moving at high speed in ether creates a bow shock at a considerable distance from the nucleus. In March 6, 1986, Vega 1, at a distance of 1.1 million kilometers from the nucleus of Halley's comet, in the direction of the sun, crossed the comet's bow shock.[56] The fact that the bow shock is created so far from the nucleus indicates that the comet must be moving through a medium.

4 - What is the cause of a comet's disintegration into smoke, dust, or ice particles? Until recently, it was thought that comets were hot, burning globes and that their tails were simply the smoke left behind. However, in 1986, this was shown to be otherwise, when several spacecraft were sent towards Halley's comet. At the closest distance of 600 kilometers to the comet made by Giotto, belonging to the European Space Agency, it was found that the nucleus of the comet is darker than expected, showing that it was not

---

[55] - The bow shock that a comet's motion makes in ether has the shape of a hyperboloid. Inside this turbulent boundary, the ether is pushed to the sides and the solar wind is decelerated quickly and then deflected around the comet's coma.

[56] - J. Kelly Beatty & Andrew Chaikin, Editors, *The New Solar System*, Sky Publishing Corporation, Cambridge University Press, Cambridge, (1990), p. 210.

burning hot. For this reason, the theory took a 180-degree turn from a hot burning fire ball to a frozen ice ball theory, which states that a comet is a dirty snowball and its tail is made from ice particles which evaporate into space by the heat of the sun.

5 - If the comet's tail is created by the disintegration of the comet, then why is no trace of the comet's tail left behind? No trace of vapour, no sign of fragments have ever been observed. In 1910, something happened which provides a clear proof that the tail of comets is not composed of gases, ice particles or fragments of the comet. In that year, the earth crossed the tail of Halley's comet, and contrary to expectations, there were no signs of any particle or gas, no observable effects[57] at all.

6 - Calculations show that for a comet the size of Halley's comet to have such a large and spectacular appearance in the sky, including a large coma and long tail, it should lose at least 25 million tons of mass every second. Such a large amount of mass, no matter what it contains, whether it be gas or ions, would be visible for a long time after the comet has left the area. How could it be that all these ice particles, dust, ions or fragments vanish completely without any observable residue?

7 - According to findings of scientists during the 1986 observations by spacecraft and on the ground, it was found that the nucleus of Halley's comet is 9 miles long and 5 miles across, the shape of a potato. The rate of loss of mass as

---

[57] - Colin Ronan, *Encyclopedia of Astronomy*, Hamlyn Publishing Group Limited, 1979, London New York, Toronto, p. 104; or *Van Nostrand's Scientific Encyclopedia*, 6th ed., New York, p. 733.

given above is so large that the comet ought to lose all of its mass in less than a day. How, then, has Halley's comet endured for so many centuries?

8 - According to some theories, the great meteor showers which occur each year during the month of August are due to the earth crossing the path of a comet. The meteors are explained as bits and pieces left by the comet striking the atmosphere of the earth. If this theory is true, why then, in 1910, when the earth crossed through the bright tail of Halley's comet, did no meteor shower occur? If the theory were true, there should have occurred one of the most spectacular meteor showers humanity has ever witnessed.

9 - The picture taken in the proximity of Halley's comet further suggests that it is not generating gas or particles, because its nucleus is clearly visible.[58] If the head and tail of the comet were made of gas and ice particles, then the photos taken hundreds of kilometers from the nucleus should show only a dust cloud or ice particles. The nucleus should be absolutely invisible, obscured by dust clouds or ice particles.

10 - If the theory were true that the head and tail of comets are caused by particles leaving the nucleus, then comets should appear with a sunlit side and a shadowy side, because the side of the comet exposed to sunlight would be very bright, and the side away from the sun would be very dark.

---

[58] - J. Kelly Beatty & Andrew Chaikin, Editors, *The New Solar System*, 3rd edition, Sky Publishing Corporation, Cambridge University Press, Cambridge, (1990), p. 225.

11 - Finally, these theories do not explain why a comet's motion creates magnetic waves or magnetic perturbation. How could the emission of ice particles, ions, smoke, dust, or bits and pieces create magnetic waves, especially if space is assumed to be free of ether? In contrast, we see that the existence of ether explains everything. In the example of the fast motion of an object in water, not only high frequency vibrations (sound waves) are generated that propagate quickly in water, but the moving object also generates slow moving waves with a high amplitude in water. The velocity of these waves is only a fraction of the velocity of a sound wave in water. We all have seen a fast motor boat passing by from a distance in calm waters. Long after the boat has passed by and gone, big shock waves reach us. Similarly, the motion of comets, apart from generating light waves, also create shock waves or etheric waves which move slowly and are called 'magnetic waves'.

When one considers that the tails of comets are only excited ether along the path of the comets, then one can see why the tail end of comets are so easily swept away from the sun by the solar wind. Since the solar wind is essentially an etheric wind,[59] the direction of the tail of the comet is simply an addition of two vectors: the speed and direction of the comet plus the speed and direction of the solar wind. If the velocity of the solar wind is much higher than that of the comet, the tail end of a comet passing by the sun will point away from the sun. This is exactly what is observed. This can also be demonstrated by an object moving in water where there is a strong current. One will notice that the tail wave of a motor boat is bent by water running perpendicular to its direction.

---

[59] - In the section called "The Source of the Solar Wind", this subject was discussed in detail.

The encounter of the solar wind with the tail of a comet also explains why a comet's tail occasionally separates from the comet. If you have run a motor boat in a strong current, probably you have noticed many occasions where the tail waves are separated by the current moving in different directions. Unfortunately, scientists have come up with a "theory" to explain this, without mentioning ether. The theory states that the solar wind has magnetic field lines; that somehow the magnetic fields of the sun are frozen in the solar wind; and that when these magnetic lines encounter the comet, they somehow, for an unknown reason, wrap around the comet. At the point of reversal, the lines pinch off ahead of the nucleus and break off from the comet. The theory, however, fails to account for the bow shock, or the coma.

As the reader will find, there are many problems with current theories. On the contrary, when we consider a comet's motion in ether, all these questions are answered with simplicity and consistency. For example, since nothing leaves the comet, no trace should be expected. Ether explains the long life of comets. Ether also explains why comets have a bright bow shock, why they have no sunlit or shadowy side. Ether explains why, on some occasions, a comet's tail points away from the sun and why comets create magnetic shock waves that reach the earth. Ether explains why the nucleus of a comet is not covered by clouds, and why we can see the nucleus even from the earth. Ether explains why, during the earth's crossing through the tail of Halley's comet, no spectacular meteor shower occurred. Ether explaine why, the width of the tail uniformly expands as the the distane from its nucleus increases.

Chapter 20

Planetary Rings

This chapter will examine how the rings of the planets further substantiate the existence of ether.

The presence of ether in space explains why there are planetary rings. Each ring is created by the motion of a fast-orbiting moon in ether. The trail of an orbiting moon is similar to a comet. Both are disturbances in ether. Here, the moon has caught up with its tail and, as a result, a bright ring, such as the one around the planet Uranus, is formed. The reason a ring around a planet appears to be of a homogeneous brightness is because the moon travels the same path over and over. As a result, a very homogeneous ring is created and the brightness enhanced. The size of the moon could be small, but the coma and the thick ring that it creates hides the moon in the ring.

One might question how the tail created by the moon could be so long as to extend around a planet. When one studies a comet, one realises that the length of the tail depends on its speed; the higher the speed, the longer the tail. For example, the tail of the Great Comet of 1843 was so large, hundreds of millions of kilometres, that it extended over one-quarter of the sky. The diameters of rings created by moons, in comparison with the tail of the Great Comet of 1843, are very small.

This understanding of the cause of the rings around planets leads us to conclude that there must be slower-moving moons that have not caught up with their tails. This is indeed exactly what astronomers have found. Observers led by Andre Brahic and William Hubbard, looking for rings around the Planet Neptune, were astonished to find that instead of a complete ring encircling the planet, something similar to a comet was found moving around the planet, with many of the characteristics of ordinary rings, but with a length that was only ten percent of the orbital circumference.[60] (Fig. 1)

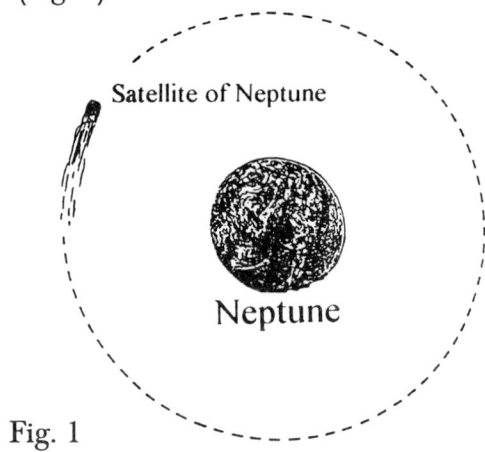

Fig. 1

The simple understanding that a ring is created by a moon explains why the rings around planets are so common.

Looking at the pictures of Saturn taken by Voyager II at a distance of 8.9 million kilometers,[61] they show three distinct rings. First of all, the mere identifiable division among rings would indicate that each ring is the path of a particular moon. Secondly, the rings are situated in one plane, being perpendicular to the axis of spin of Saturn. This further establishes that the rings are the paths of different moons. Comparing the paths of the moons of Saturn to those belonging to Jupiter, we see that the paths of the moons of Jupiter are situated in one plane. We must bear in mind that the paths of all the planets revolving around the Sun lie more

---

[60] - J. Kelly Beatty & Andrew Chaikin, Editors, *The New Solar System*, Sky Publishing Corporation, Cambridge University Press, Cambridge, (1990), p. 162.

[61] - Ibid, p. 163.

or less in one plane. Fig. 2 shows the similarity of the path of Saturn's moons with those of the planets around the sun.

Fig. 2

Since the light of each ring is from the friction and vibration that the motion of each moon generates in ether, then the speed of each moon must have a direct relationship with the colour of the ring. The higher the speed of a moon, the more bluish or brighter the ring becomes. The lower the speed, the more reddish or lighter the colour of the ring. Since the moons have different orbital velocities, the colour of the rings must also be different. Looking at the images of the rings of the planet Saturn, we see these precise differences in the colours of the rings.

Fig. 3

Looking at the shape of the rings of Saturn in Fig. 3, one can notice that they are flat with thin edges. The shape of Saturn's rings leads us to infer that around each of the moons of Saturn must revolve a small satellite. The existence of a relatively small satellite, moving so fast that it can orbit its moon many times before the moon completes a revolution, explains why the rings are flat. The fast motion of a satellite, orbiting both a moon and Saturn at the same time, creates a wide and flat disk as seen around Saturn. To see the reason, imagine a satellite moving very fast around a moon. If the moon were stationary, the tail wave of the satellite would form a "hulahoop"-style ring around the moon as shown in Figure 4.

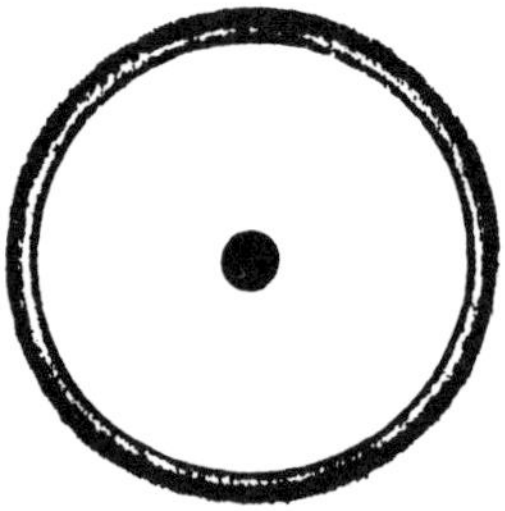

Fig. 4

Since the moon is not stationary but revolves around Saturn and carries the satellite around, the tail wave of the satellite is not a thin hulahoop ring, but a wide and flat ring as shown in Fig. 5, because the hulahoop ring is dragged in its plane, all around the planet.

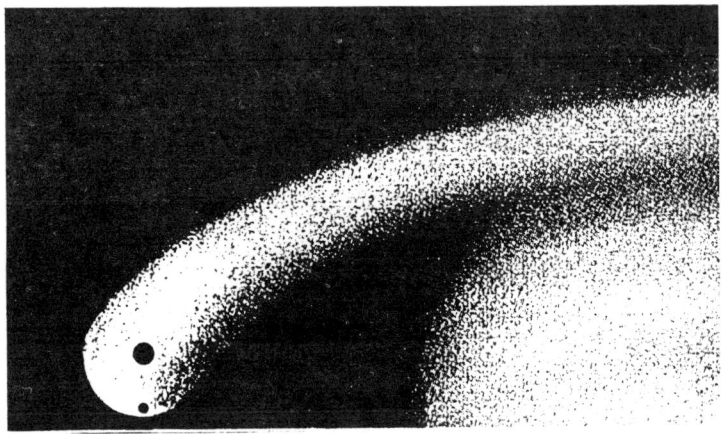

Fig. 5

The width of the ring must therefore be equal to the diameter of the satellite's orbit around the moon. We must bear in mind that as the path of a satellite, orbiting both the moon and the planet, lies in one plane, the ring that it creates must be flat and have a relatively thin edge. This is exactly the shape of the rings of Saturn. Furthermore, looking at Figure 6, one can see that the speed of the satellite at point D is slower than at point C, because at point D, the direction of motion of the satellite is opposite to that of the moon, but at point C, the direction is the same. At point D, the velocities subtract; whereas at point C, they add up.

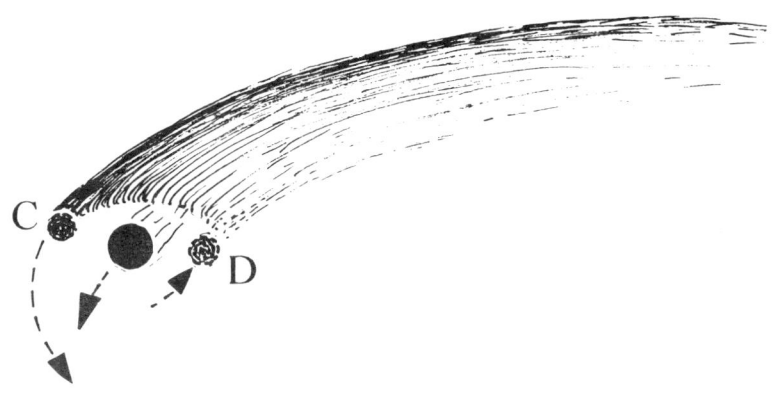

Fig. 6

This shows that the satellite at the inner edge of the ring (at point D), where it is closest to Saturn, has a net velocity different from when the satellite is at the outer edge of the ring (point C). This shows that the colour of the inner edge of the rings must be different from the outer edge. This, again, is exactly what the images show, and is exactly what scientists discovered by a method called "red shift effect". The first ring in Fig. 3 shows that, at the inner edge of the ring where the net velocity of the satellite is lowest, the colour is lighter than at the outer edge, where the net velocity is highest. Fig. 7 shows that the direction of the orbit of the satellite is reversed. Notice that at the inner edge of the ring, where the net velocity is highest, the colour is brighter than at the outer edge, where the net velocity is lowest. See the second ring in figure 3.

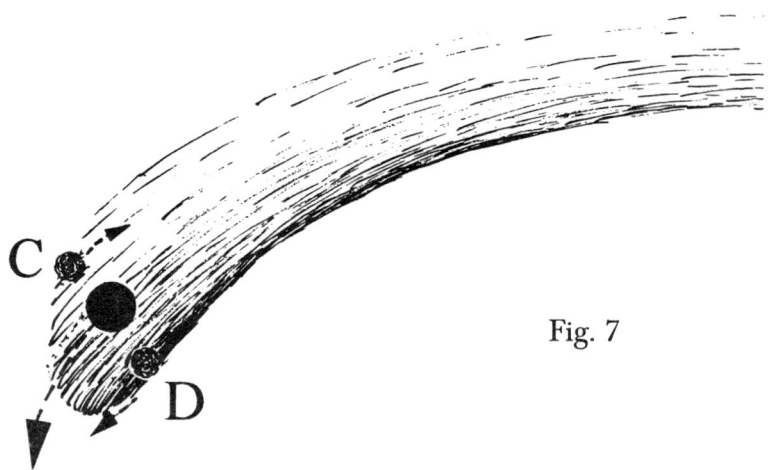

Fig. 7

The fast orbital motion of a satellite around a moon not only generates friction and light in ether, but also creates waves in ether. The waves generated by each of the satellites belonging to different rings interfere with each other and create standing waves that appear to us as many ringlets. This is, again, exactly what scientists have found. Pictures taken of the rings of Saturn show many wave pattern ringlets within the rings, as in Fig. 8.

Fig 8

According to some theories, the rings are made of ice particles, or bits and pieces of rock that have left the moon, breaking up entirely as they travelled too close to the planet. None of these theories explains how it is possible for a moon to break up in such an orderly fashion and create different colours of rings. Furthermore, how is it possible for the moon to disintegrate into ice particles, ions or dust, allocating different colours of ice or dust to different rings? How is it possible for a moon to disintegrate in such an orderly way such that nothing has been observed between the rings? None of these theories can explain the stability of the rings. If we assume that the rings are made of ice particles or dust, then in a relatively short period of time, the ice particles or dust in the rings should have mixed with that of other rings, and spread over and covered the entire atmosphere of the planet. Why have the rings of the planet persisted unchanged for so long? The presence of moons explains the stability of the rings without any difficulties, as the presence of the moon revolving about the earth has been stable and lasting. The theories do not explain why the brightness and colour of the inner edges of the rings differs from that of the outer edges. Finally, the theories do not explain why there are many standing waves in the rings.

Chapter 21

The Motion of Asteroids in Ether

The following is an account of a phenomenon that I myself observed. This phenomenon provides a very clear evidence that ether exists, and that the bow shocks of comets and that of the earth are all created by their motion in ether.

One night, while I was watching the stars,[62] I suddenly noticed a bright object that appeared to be a satellite. I later realised it was an asteroid moving in space. As it moved, I saw that the light of the stars nearby was momentarily affected. As the asteroid moved, the light of the stars located on both sides of its path either oscillated back and forth or was momentarily interrupted. It was as if I were watching the image of the stars in clear water and a small wave, such as from a tossed pebble, shook the image and interrupted the starlight, in the twinkling of an eye. From the way the starlight was affected, it became very clear that the etheric shock waves created by the motion of the asteroid were responsible for the phenomenon.

This phenomenon has surely been seen by many others. I cannot believe I am the only witness. It is one of numerous evidences of the existence of ether by direct observation.

---

[62] - It is very interesting to note that this incident happened while I was writing the section about the Magnetosphere of the Earth and comets.

Chapter 22

The Motion of the Earth in Ether

Apart from the earth's magnetic field, scientists have discovered that the earth is the centre of huge, invisible magnetic waves that have the overall configuration of a comet. Fig. 1 shows the earth as the nucleus of a magnetic bow shock and a long magnetic tail wave.

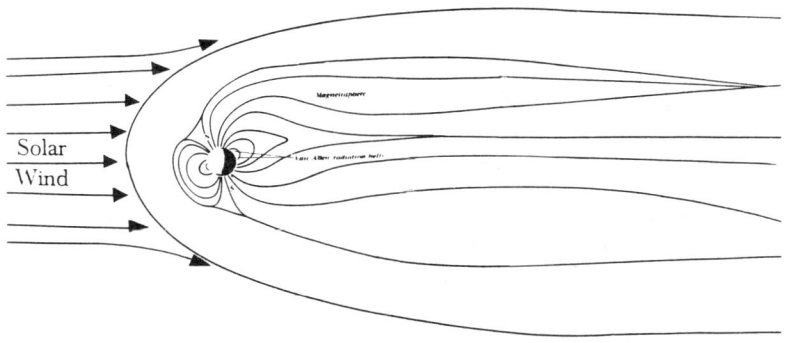

Fig. 1

Unfortunately, Einstein's followers have not related this phenomenon to ether and to comets, in order to consider that the cause could be the same.

The motion of the earth in ether is similar to the motion of comets in ether, but because the earth's motion is relatively slow and does not move against the solar wind, the waves are not luminous. The earth moves so slowly that its etheric shock waves are not strong enough to create sufficient friction to generate light, nor does the encounter of these shock waves with the solar wind create light. For these reasons, the overall magnetic waves do not glow as do comets.

It was mentioned before that the motion of the earth in ether is very similar to the slow motion of a spherical object in water. Fig. 2 is a picture taken of the waves that the motion of a ball has created on water. The floating ball was pulled very slowly by a string. Notice how the shapes of the shock waves created in water are similar to those in Fig. 1. The bow shock and the tail waves are exactly similar to those created by the earth's motion in ether.

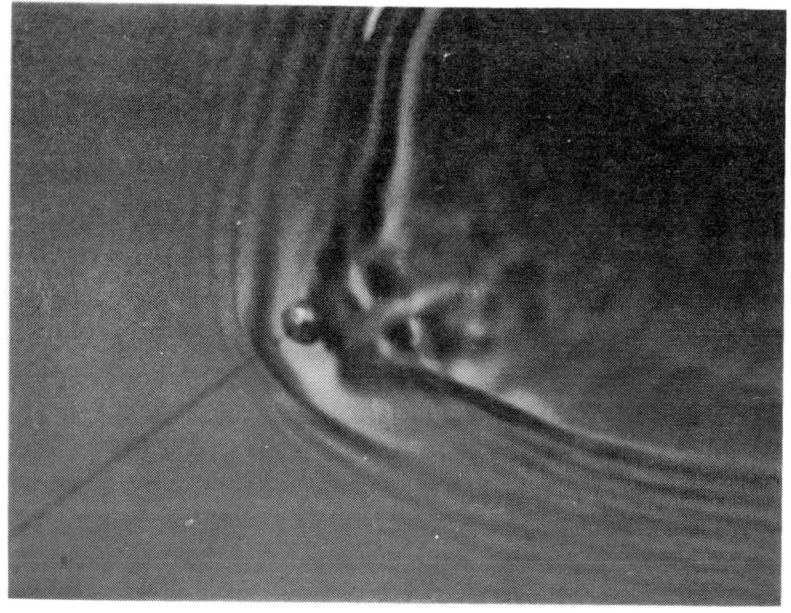

Fig. 2

Scientists have found that the bow shock exists at a relatively very long distance from the earth. The solar wind, which moves with a speed of at least 400 kilometers per second, is suddenly slowed down and diverted to the sides by the bow shock at a distance of about 10,000 kilometers from the earth. Relativity defenders claim that the existence of the bow shock is due to the encounter of the solar wind with the magnetic field of the earth. However, the shape of the bow shock and its far distance from the earth gives a clear indication that the cause of the bow shock is not a magnetic field, but rather, the motion of the earth in ether. If one looks at the figure scientists have provided, one sees that the magnetosphere of the earth is distinct from the bow shock. The magnetosphere surrounds the magnetic field. One can see also that the magnetosphere is the limit of the earth's magnetic field, and is located at a far distance from the bow shock. This clearly shows that the cause of the bow shock could not be the encounter of the solar wind with the earth's magnetic field. Although the bow shock shields the earth from the solar wind, its existence is due to the motion of the earth in ether.

The reason that the tail wave is in a direction away from the sun is exactly the same as for that of comets. We know that the tails of slow-moving comets near the sun always point away from the sun. This is because the solar wind moves much faster than the comet. Bearing in mind that the solar wind is in reality the wind of ether,[63] then the direction of the tail wave is simply the addition of two vectors: the speed and direction of the comet plus the speed and direction of the solar wind. Since the velocity of the earth is much slower than the solar wind, the tail wave is in a direction away from the sun. This again can be clearly illustrated by an object moving in running water.

---

[63] -See Chapter 25, "The Source of the Solar Wind".

Chapter 23

The Aurora Borealis

The Aurora Borealis, commonly called the Northern Lights, is a flickering natural light that occurs high above the atmosphere and is seen at night in the sky of the northern hemisphere. A similar light, Aurora Australis, is seen in the southern hemisphere. It has been found that when a magnetic storm created by sunspot activity reaches the magnetic field of the earth, it creates the Aurora displays. It has also been found that when magnetic storms are very strong, the aurora displays are not only brighter and more spectacular, but they extend far toward the equator. Scientists believe that the auroras have something to do with the interaction between the solar wind and the earth's magnetic field. This is because they have found that the higher the speed of the solar wind, the brighter and more spectacular the auroras become. However, the exact reason why the encounter of the solar wind or magnetic storms with the magnetic field of the earth creates the flickering of light has remained unknown.

With the understanding that the solar wind, the magnetic storms and the magnetic field of the earth all in reality arise from the current of ether, we then can see why the encounter of the winds moving in different directions can create friction and vibrations in ether. Some of these vibrations are in the visual frequency range which we see as the light of the aurora. Since at the poles the magnetic pressure is highest, and some subatomic particles are trapped, the encounter with the solar wind generates the greatest friction. For this reason, at the poles the aurora displays are more frequent than at other places. From the fact that the light of auroras is from friction in ether, and that the light of comets also is from friction in ether, one can see that they are in fact the same phenomenon. This is especially demonstrated in the period of greatest sunspot activity, when aurora displays extend far towards the equator. This shows that if the magnetic storms were much stronger, then the whole of the sky over the earth would glow just like the coma of comets. Furthermore, the earth's magnetic tail wave, created by the motion of the earth in ether, would also glow like that of a comet.

Chapter 24

The Sun As A Source of Light

This chapter suggests that the sun as a source of light provides evidence in support of the existence of ether.

There is a common theory that explains the sun as a glowing ball of hot gases and that the entire mass of the sun is composed mainly of hydrogen and helium atoms.[64] It is believed that billions of years ago, "clouds of interstellar gas and dust, perhaps from an explosion of a nearby star, collapsed and condensed to form the solar system".[65] Theoreticians claim that the light and heat of the sun are from nuclear reactions which, for unknown reasons, allegedly occur at a very controlled and continuous rate, so that the sun neither heats and explodes nor cools too rapidly. It is further believed that eventually the nuclear reactions will exhaust themselves, resulting in the cooling and darkening of the sun. They claim that the sun, after losing much of its energy, one day will expand and its volume become billions of times bigger. Then its atoms will collapse upon themselves and the sun will shrink into a ball of a few miles in radius. They also claim that the enormous gravity of the sun at the present time is the result of the collapse of hydrogen atoms stripped of their electrons, thus bringing the atoms closer together and creating a much greater density.

In this chapter, it will reasoned that most of the above theories regarding the sun are erroneous. Then it will

---

[64] - John Gribbin, "Inside The Sun", *New Scientist*, (Inside Science), No. 60, (March, 13, 1993) v. 137, p. 1.

[65] - J. Kelly Beatty & Andrew Chaikin, Editors, *The New Solar System*, Sky Publishing Corporation, Cambridge University Press, Cambridge, (1990), p. 1.

be reasoned how light and heat are generated on the sun through the effects of ether.

Theoreticians say, *"A star is formed when a large amount of gas (mostly hydrogen) starts to collapse in on itself due to its gravitational attraction."*[66] If we think about this theory, we see that it is based on impossible assumptions, because hydrogen gas does not have much mass or gravity to cause the collapse of the gas upon itself.

Let us assume that the theory is true and the sun is composed of hydrogen gas. Then we can ask, how did such an immense amount of gas come together in the first place, especially in such a highly compressed form? Obviously, the gas must have been under tremendous pressure from the outside and from all directions, leaving no way out for the gas to escape in order to be compressed to a degree that its atoms lose all their electrons. The pressure could not have come from within, because a gas, especially hydrogen gas, does not have much mass or gravity to create such an immense force of compression. The force could not have come from the explosion of a nearby star, as a few theoreticians have suggested, because an explosion either creates a sphere of a plane wave that pushes the gas away from the centre, or it will send fragments to pass through the gas. Where, then, did the force of compression or attraction come from?

Moreover, the sun is not the only star in the universe. If the sun is the result of the compression or attraction of gas by unknown forces which somehow applied pressure from all directions, then all other stars must be formed by similar forces. Where did these forces come from?

---

[66] - Stephen W. Hawking, *A Brief History of Time*, Bantam Books, Toronto, New York,(1988), p. 82.

In contrast with the above theory, consider the fact that atoms attract each other, and as atoms combine, gravitational force increases. For this reason, a small amount of mass can gradually attract dust or other masses, in a process called the 'snow ball effect', to form very large universal bodies like the sun. This understanding is based on a natural and universal law that is so simple that it needs no further explanation.

The reason that led scientists to conclude that the sun is mainly composed of hydrogen gas is because they have found that 97.5% of the gases covering the sun are hydrogen and helium. The remaining 2.5% are nitrogen and other gases. Scientists have merely discovered the composition of the outer atmosphere of the sun. What has been seen of the sun from far distant earth is only the outer atmosphere and not deep below the gases and clouds that cover the actual surface, because the thickness of the outer and inner layers of the atmosphere of the sun are each hundreds of times thicker than that of the earth. It is very difficult to analyse the composition of the actual surface located deep below the masses of gas and clouds. There, the high gravitational force, along with the very high elevation of gas, together create an immense atmospheric pressure that compresses the gas to an extremely high density that can hide the actual surface completely. The composition of the outer atmosphere of the sun is very similar to our own planet. If the composition of the earth were analysed from a great distance away, we probably would have made the same conclusion as did the scientists on earth studying the sun. After all, is not the outer atmosphere of earth also made up of hydrogen, helium and small amounts of nitrogen?[67]

---

[67] - Unfortunately, scientists have made similar mistakes about the composition of the Planet Jupiter.

Telescopic photographs of the sun show the presence of clouds in its inner atmosphere. This configuration is also very similar to our planet, as if a similar picture were taken by the satellite showing the clouds over the ground (Fig. 1).

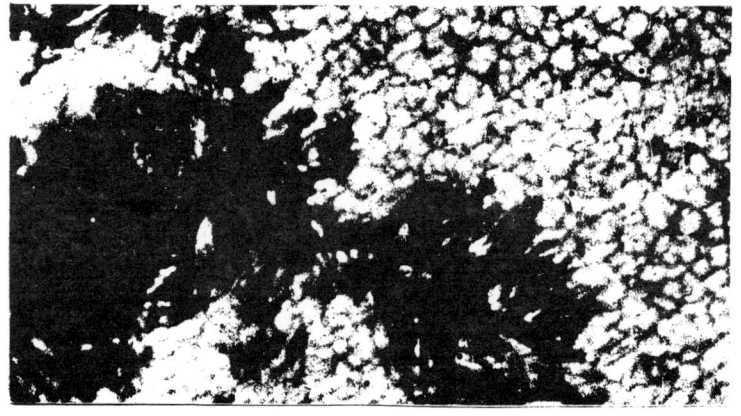

Fig. 1

This shows that the conclusion scientists have made about the actual composition of the sun is incorrect. The following are some suggestions to trigger new thinking regarding the sun:

1 - The presence of radioactive materials reasonably accounts for the gravity of the sun without the hypothesis of a composition of collapsed atoms of hydrogen to make it denser and heavier.

2 - The presence of radioactive elements[68] also accounts for the light and heat of the sun without the hypothesis of nuclear reactions within the sun.

The continuous formation of highly charged ions by the presence of radioactive elements below the atmosphere can produce extremely powerful winds and friction that affect ether and create heat and light with very high intensity. On earth there are always winds. There are always a few storms and lightning. There is always friction created by winds, but the density of air is very low and the winds are not strong enough to affect ether and generate light and heat. On earth, relatively few storms and lightning occur, whereas on the sun, the case is quite different. There are always numerous storms, each having so much energy and power that nothing on earth can be compared to them. The storms are so powerful that they throw ether into extreme excitation and vibration. The vibration of ether comes to us as the light and heat of the sun. The continuous ionisation of gases by radioactive elements and the surge of ions from one place to another can be proven by the fact that researchers from the California Institute of Technology have discovered that all over the surface of the sun, small patches move in and out 10 to 12 times per hour with velocities of about 500 meters a second and an overall displacement of 50 kilometers.[69] All this shows that below the apparent surface there must be numerous surges of gas from one location to another going

---

[68] The author strongly believes that there exist on earth a series of rare and as yet undiscovered elements that have qualities opposite to that of radioactive elements. These elements instead of emitting radiation, steal radiation. The atoms of these elements are smaller than that of the hydrogen atom. The author suspects that the presence of these elements in the human body is the cause of cancer.

[69] - John Gribbin, "Inside the Sun", *New Scientist*, No. 60, v. 137, (March 13, 1993), p. 4.

on all over the sun, each adding to the turmoil, putting the ether into vibration and creating light and heat and even radio waves. Here we see why the sun has been an almost inexhaustible source of light and heat. The sun gives energy as long as radioactive elements last.

The light and heat of the sun are simply from very strong storms that affect ether rather than from unexplained, unnatural, unstable, uncontrollable, and rapidly exhaustible nuclear reactions of hypothetically collapsed hydrogen atoms. These storms provide a natural, continuous, stable and inexhaustible source of light and heat.

3 - The lightning caused by the clashes of oppositely charged clouds solves one of the mysteries in radio astronomy.[70] Everyone must have noticed that during thunder and lightning storms, we hear little bursts of static on the radio. The clashes of oppositely charged clouds create disturbances in ether and generate light waves as well as radio waves and other waves. The burst of radio waves coming from the sun is also from the clashes of highly charged clouds, which create intense waves, vibrations, and winds of ether surrounding the sun. These waves and vibrations reach us as the light, heat and radio waves of the sun, and the etheric winds may be strong enough to reach us as a magnetic storm.

Furthermore, considering that the sun is composed of radioactive elements, we then are led to the understanding that below the atmosphere there must exist a sea of liquid that enhances the storms. This is because the heavier atoms stay below the lighter ones. Since the thickness of the atmosphere of the sun is at least a hundred times greater

---

[70] - Patrick Moore, *The Sun*, Frederick Muller Limited, Great Britain, (1968), p. 88

than that of the earth, this alone can create a very high atmospheric pressure. Considering that the sun has a very strong gravitational force, and that deep below the gases, the gravitational force is even stronger, then the atmospheric pressure must be so immense that the atoms are compressed to such a degree that any gas becomes liquid.

Although scientists tell us that the sun is composed of gas, what they actually are implying is that the sun is composed of liquid or even solid. They explain that the sun's hydrogen atoms, being under immense pressure and heat, have lost their electrons and the nuclei have come very close to each other. This means that the main composition of the sun must either be liquid or solid and not a gas, because first of all, we know that hydrogen gas under pressure converts into a liquid. Secondly, is it not the electrons that change a solid into liquid or a liquid into a gas? Moreover, the temperature above the apparent surface of the sun has been found to be rather low, "less than 4000 C".[71] The American Encyclopaedia writes that:

> ... *above the apparent surface the temperature first decreases to about 7200 F or 4000 C and then increases rapidly to more than 1,800,000 F (1,000,000 C) in the outer atmosphere of the sun.*

The apparent surface temperature decreases to 4000 C, indicating that below the apparent surface, the temperature must continue to decrease to much lower temperatures, a condition suitable for the presence of liquids. All this means that the entire surface of the sun must be

---

[71] - John Gribbin, "Inside the Sun", *New Scientist*, No. 60, v. 137, (March 13, 1993), p. 4.

covered by a limitless sea of liquid. The presence of the sea and the gas above it explains why scientists have not been able to determine the exact period of rotation of the sun.

Considering the vastness of the sea and its powerful storms, giant waves of liquid are successively created that move with a very high speed over the sea. These waves in turn create intense waves and vibrations in ether which is covering the sea. These waves and vibrations are transmitted by the medium of ether in space and reach us as the sun's light and heat.

An example will illustrate how light is created by the sun. On a windy day when we stand on the shore of a turbulent sea, we can hear the very loud and continuous sound of the waves. In this case, the wind, along with the waves of water, causes the sound from the sea. All these show that as the turbulent seas generate sound, the turbulent sea of the sun creates light. The sea that covers the sun is so vast·and great that it is at least twenty thousand times greater than the Pacific Ocean. Hence, it is very logical to assume that such a vast and deep ocean of heavy liquid of high radioactivity, and with a high temperature, creates extremely powerful storms. Furthermore, such a great ocean of hot and heavy liquid must have numerous volcanic and gas eruptions, like a boiling pot of liquid, some of which are gigantic and powerful, millions of times greater than the Mount St. Helen eruption. Therefore, the storms created must be millions of times greater and more violent than that of a storm in the Pacific Ocean. Such a violent ocean must create giant waves, hundreds or even thousands of kilometers in height, moving with incredibly high speeds, the top front section of each wave pouring down by the high gravitational force of the sun and roaring as a super giant waterfall, creating light and heat with incredible intensity and power. It is therefore proposed here that the light of the sun results only from the excitation of the medium of ether that surrounds the sun, and not from nuclear reactions.

In contrast with this view, which is so simple and logical, it has been suggested that the light and heat of the sun are from nuclear reactions that, for unknown reasons, occur deep inside the sun. If this theory is true, no sunspots, the large dark spots, should appear on the surface of the sun. Instead, the whole surface of the sun should always give light as the heat from inside the sun would cause the whole globe to glow. Moreover, it is assumed, again for unknown reasons, that these nuclear reactions proceed at a controlled rate, and that the sun is neither heating nor cooling too rapidly. What regulates the reactions? This theory has also failed to answer simple questions such as why, if nuclear reactions occur deep inside the sun, are the temperatures below the atmosphere much lower than high above the atmosphere? Furthermore, this theory has not been able to explain the reasons for radio waves, magnetic storms, etc.

Here we see that the light and heat of the sun are only excitations resulting from friction in ether, which is greatly enhanced by the clouds above the sea. The light and heat represent the great magnitude of turmoil that exists on the sun.

The phenomenon of the solar flare represents a violent storm on the surface of the sun. According to present theories, what we see of the solar flare is the result of an enormous amount of gas shooting out with a speed close to the speed of light, travelling millions of miles above the sun's surface. The theories also assert that these gases, despite the high gravitational force, leave the atmosphere of the sun and that somehow the magnetic field of the sun becomes frozen inside them. Occasional flare-ups of such activities create magnetic storms on the surface of the earth.

If these assumptions are true, then why, during an eclipse of the sun, has it been found that the space surrounding the sun is a "perfect vacuum" without the trace of any gases? Furthermore, what does the release of gases have to do with magnetic storms on the earth? How could a gas change into a magnetic storm?[72] Why does the magnetic storm reach the earth's surface without the gas?

The cause of the magnetic storm could be gigantic volcanic or gas eruptions on the hot ocean of liquid. The high gravitational force, the heavy atomic weights of the elements, the depth of the sea, the high temperature, are all ingredients that can produce eruptions with such great power and proportions, eruptions that are billions of times greater than any volcanic eruption on earth. Naturally, such powerful eruptions must produce tremendous pressure in ether and create very strong winds of ether that reach us as magnetic storms. That is why the magnetic storms are sometimes experienced after seeing solar flares. A magnetic storm is, in reality, an ether wind, which moves much more slowly than a light wave, just as air winds move much more slowly than sound waves.

---

[72] - There is a theory based on many hypotheses which assumes that the magnetic field of the sun is somehow frozen by the gases and that when these gases reach the earth, the magnetic field is separated from the gases and reaches us as magnetic storms.

Chapter 25

The Source of the Solar Wind

The solar wind is yet another phenomenon that supports the existence of ether. The existence of ether in space and the spin of the sun in ether are the causes of the wind. In other words, the spin of the sun in ether creates what we call the solar wind.

There are different theories about how the solar wind is generated. They are based on one common assumption: that the solar wind is generated by gases leaving the sun. It is suggested that there is a continuous flow of gas from the atmosphere of the sun into the solar system. As a result, the sun is continuously losing 1,000,000,000 kilograms of its mass every second[73] to make up the solar wind.

This chapter will provide clear reasons showing that nothing leaves the sun, no particles and no gases leave the atmosphere, but rather the rotation of the sun along with the ether surrounding it creates the solar wind.

Since the sun spins in a space filled with ether, the spin of the sun causes the ether about it to revolve spirally. This spiral motion of ether is what we call the "solar wind". Moreover, since ether is matter, it is therefore being attracted by the gravity of the sun. The ether together with the sun, rotates about one axis, just like the earth and its air atmosphere rotate about one axis. As the distance from the sun increases, the influence of the gravity of the sun decreases. The circular motion of ether by the spin of the sun, together with the centrifugal force, create a wind moving in an expanding spiral direction (Fig. 1), similar to the spring of a watch.

---

[73] - Colin Ronan, *Encyclopedia of Astronomy*, Hamlyn Publishing Group Ltd., London, New York, Sydney ,Toronto, 1979, p. 56.

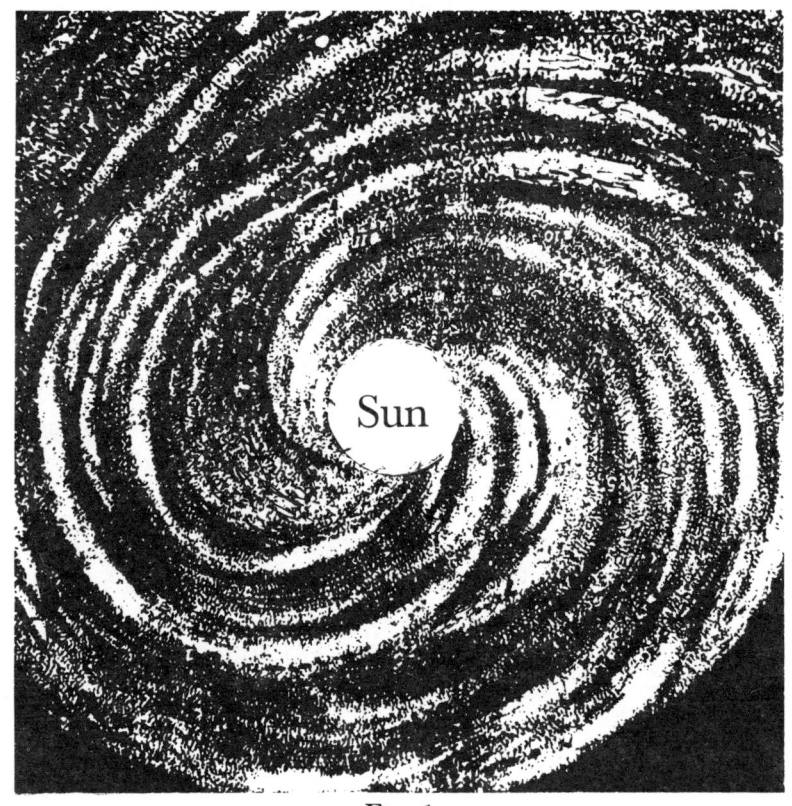

Fig. 1

Assuming the rotation of the sun causes all the ether in the solar system to rotate with the sun as if they were one rigid body, then as the distance from the sun increases, the velocity of ether also increases. Since the spin velocity of the sun at its equator has been found to be one revolution per 25 days or 4.63×10^{-7} rev./sec., and the distance from the sun to earth is known to be 1.49×10^{11} meters, then the orbital velocity of the ether must be equal to: $(2 \pi r) \times ($ spin velocity$)$ or:

2 x 3.14 x 1.49 x 10 ^{11} m x 4.63 x 10 $^{-7}$ rev./sec. = 433 km/sec.

The solar wind at the orbit of the earth has been measured by scientists[74] and different values have been found. Some have reported a velocity of about 400 kilometers per second, some reporting 500 kilometers per second, and all being very close to the above calculated value.

The spin of the sun about its axis creates two poles. At the poles where the axis of the spin passes, no centrifugal force exists, while at the equator the centrifugal force is maximum. The spin of the sun in ether creates two suction tunnels that continuously, in the direction of the axes, attracts the ether towards the poles. The ether is then pushed towards the equators and then, by centrifugal force, is pushed away from the sun in an expanding and spiral direction into the solar system. This can be demonstrated by the spin of a spherical object in water. The ether carries along its path small particles such as protons and electrons.

In 1957 Hannes Alfven discovered that the solar wind must have magnetic qualities and that the "magnetic field carried along by the solar wind plays a vital role in cometary interaction."[75] Unfortunately, scientists have not related the solar wind to ether; nor have they realised that the magnetic force is created by the flow of ether, or that the solar wind is created by the spin of the sun in ether. Scientists all agree that the solar wind has magnetic qualities and that the wind has something to do with the spin of the sun. They have even concluded that the solar wind in the solar system

---

[74] - Ibid, p. 56.

[75] - J. Kelly Beatty & Andrew Chaikin, Editors, *The New Solar System*, Sky Publishing Corporation, Cambridge University Press, Cambridge, (1990), p. 226.

moves spirally[76] (Fig. 2), but they have not related the wind to ether.

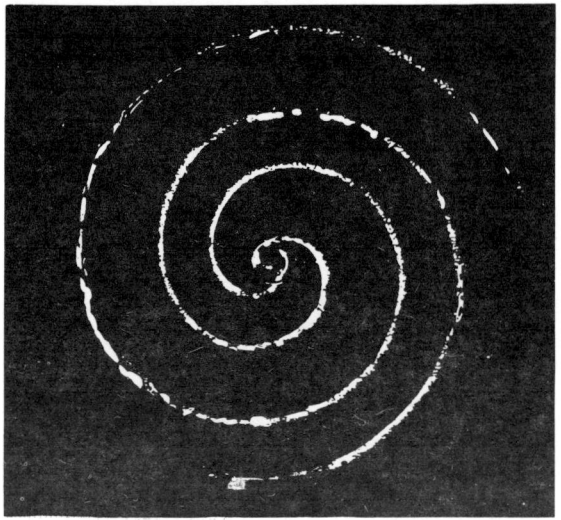

Fig. 2

Here we see that ether provides a very simple answer to all the questions related to the solar wind. By contrast, if we assume that the solar wind is generated by gases leaving the sun, then problems arise. For example, if gases are released by the sun, how can we explain the fact that during a solar eclipse, it has been found that the space immediately surrounding the sun is a perfect vacuum? If gases are leaving the sun to make up the solar wind, then near the sun the density of the gases must be high and not a vacuum. Moreover, hot gases leaving the sun would distort the light of stars considerably. We know by experience that hot air rising from a hot stove distorts the light passing above it. Similarly, hot gases leaving the sun would distort the light of stars. The

---

[76] - Ibid, p.29.

fact that there is no such distortion suggests that no gases are leaving the sun.

Finally, from where does the magnetic quality of the solar wind come? Scientists have assumed that the magnetic field of the sun is somehow frozen into the gases that leave the sun. If this idea is true, then when magnetic storms reach the earth, they must have been carried by these gases. Why, then, are magnetic storms that reach the surface of the earth not accompanied by any gas?

Chapter 26

Pulsars

In 1967 Jocelyn Bell, a student, discovered that regular pulses of radio waves were coming from an unknown object in the sky. The orthodox followers of Einstein argue that these radio waves prove the existence of hypothetical "neutron stars". They claim that the atoms of these so-called 'neutron stars' are very different from anything we have known on our planet or in our solar system. They assume that the atoms of these neutron stars do not have any electrons, but instead, have neutrons which are circling the nuclei. They also assume that a neutron star is very small, with a radius of only 10 kilometers, but that its mass is immense, equal to that of the sun. They further assume that the density of these stars is hundreds of millions of tons per cubic inch. Moreover, they assume that the spin of these stars generates radio waves. With all these fabricated assumptions, they did not solve the problem of how the spin of a star can generate a radio wave. Nevertheless, they teach what they have claimed as if they are facts.

Here it will be reasoned that pulsars are simply meteorites, asteroids or even stars, that are spinning in space. The regular pulses or radio waves are simply generated by their spin in ether. Just as fast-spinning objects in air generate sound waves, similarly the spin of stars, meteorites or asteroids in ether generates radio waves.

We know that in space there exist all kinds of meteorites and asteroids made of rock. Some of these must be spinning very fast, given that the stars, the sun and its planets, the nuclei of atoms and their electrons, are all spinning. Is it an unreasonable proposition to assume that some of the meteorites or asteroids could also be spinning?

particularly when we consider the fact that there are many reasons for their spin? Given the fact that some meteorites or asteroids are moving with very high velocities in space, there is a high degree of possibility that one of these bodies might pass so close to another that they actually touch one another, causing each other to spin rapidly. Such collisions naturally might cause them to spin. The fast spin of a meteorite or an asteroid in ether generates disturbances or waves in ether. These waves come to us as radio waves, just as the spin of an armature motor generates sound waves. Furthermore, since the meteorites or asteroids are relativity small and do not radiate light, they are not visible, even by telescopes. Since some of the asteroids are less than a hundred meters long, and some are made of very hard rock or metal, frozen to temperatures below - 150° Celsius, their spins will not break them apart.

In contrast to this simple understanding based on facts of nature, the hypotheses of the neutron stars are based on unnatural and unrealistic assumptions. It is assumed that in one cubic inch there could exist hundreds of millions of tons of mass. This assumption obviously lacks the understanding of why atoms have a limited mass; why electrons and not neutrons circle around the nucleus; and how the gravitational force is created. It lacks the understanding that the nuclei of atoms spin in ether, and that the spin in ether is the cause of the gravitational force, as well as the cause of the limited sizes of the atoms. Nuclei that are too close to each other will create friction, slowing down each others' spin velocities to the point where friction destroys their existence. Finally, and most significantly, with all these assumptions, Einstein's followers have not been able to come up with a simple reason as to how and why neutron stars generate radio waves.

When one carefully considers the assumptions, it becomes clear that all are unrealistic and unnatural, the product of pure imagination. They were made only because they were based on Einstein's idea that space is empty, without ether. When one realizes that space is not empty, but rather, is filled with ether, then all the unrealistic assumptions are avoided altogether.

Chapter 27

The Mysterious Explosion of 1908

In June 30, 1908, in the Tunguska River area of central Siberia, a brilliant fireball was seen in the sky and an immense explosion occurred. Trees were uprooted 40 kilometers away, and pressure waves were recorded as far away as the British Isles. Residents of this sparsely-populated area sustained flash burns as far away as 60 kilometers. While the event showed all the signs of the fall of a large meteorite, no sign of impact or any major fragments of the meteorite have ever been found. Recordings of fluctuations in the earth's magnetism at the time showed an effect strikingly like that produced by an atomic bomb explosion in the atmosphere.

Orthodox believers in Einstein have advanced several theories about this event. Such theories include a black hole falling on the earth, an anti-rock made of antimatter falling and exploding, a small comet largely composed of ice and solid particles falling and exploding in mid-air, and an advanced civilisation setting off a nuclear explosion. Another theory suggests that a damaged flying saucer from outer space may have caused the explosion.[77]

None of the theories could provide the answers to all the questions related to the event. For this reason, its cause has remained a mystery. To understand what happened, let us review the phenomena of comets and nuclear bomb explosions.

In Chapter 4, "Nuclear Explosion vs. Non-Nuclear Explosion", it was mentioned that in a nuclear explosion, it is the ether shock wave and its pressure which causes immense

---

[77] - *The World's Last Mysteries*, The Reader's Digest Association Inc., New York, Montreal, 1978, p. 295.

destruction and creates light, heat and other forms of radiation. In Chapter 19, "Comets and Ether", it was mentioned that the motion of an object in ether creates both shock waves and a coma which is far larger than the size of the object. For example, looking at a picture of a comet, one can see that the size of the coma of a comet is much larger than the size of its nucleus. With these facts in mind, we can solve the mystery.

What happened in Siberia was the fall of a small meteorite. The destructive power was from the speed, size and power of its shock wave reaching the atmosphere, and not the size of the meteorite itself. Although the meteorite was small, its very high velocity, (which would have been tens of thousands of kilometers per second), created a powerful etheric shock wave which caused all the destruction. This is why no large crater or major fragments were found. This is how the meteorite created magnetic shock waves similar to a nuclear explosion. This is why the shock wave created the various radiations, including light and heat.

Imagine a small meteorite, 50 centimeters in diameter, but with a coma shock wave of a few hundred meters in diameter, entering the atmosphere with a speed of 25,000 kilometers per second. The shock waves accompanying the meteorite would have the same characteristics as those of a nuclear bomb explosion. This kind of shock wave has a tremendous destructive power that can blow trees down at considerable distances, as the force of the shock wave is much greater than the attractive force binding the atoms of the trees (bearing in mind that all molecular bindings arise from the force of ether). The magnetic wave recorded was caused by the pressure of ether. As the meteorite entered the atmosphere, all the ether in its path was rapidly and with great force pushed aside. Since the magnetic phenomenon is created by the flow of ether, this is what was recorded. Of course, the column of light observed is explained by the fact that a fast-moving object in ether creates a long and bright tail like a comet.

Chapter 28

The Earth Drags Ether

The nineteenth-century scientists, prior to Einstein had concluded that the earth drags ether, by studying the phenomenon of stellar aberration and through experiments such as that of Michelson. Einstein altered this understanding, and with it changed the course of science. This will be discussed more fully later. Here, evidence that the earth drags ether will be detailed from the following six perspectives:

1. The earth's drag of air atmosphere must accompany ether
2. The comets
3. Discovery of the magnetic bow shock and the solar wind
4. The earth's magnetic field
5. Michelson's experiment
6. Stellar aberration

1 - We know that the earth drags its air atmosphere. This in itself proves that the earth must also drag the ether particles that fill the spaces between the atoms and molecules of air. To think that the earth drags along all the atoms and molecules that make up its atmosphere yet leaves out the ether particles is absurd, bearing in mind that all the atoms and molecules are made of ether particles. The idea that the earth drags its air atmosphere without ether would imply that all the atoms and molecules of air are constantly encountering a 30 kilometers per second wind without being affected by it. Obviously, this idea could not be true.

2 - In previous chapters, we saw how the earth's motion in ether has some similarities to that of a comet in

ether. Looking at the figure of a comet (Fig. 1), one can see that the nucleus is small, but around it there is a large and bright head called the 'coma'. The coma is, in reality, the ether that the nucleus drags, which has become luminous.

Fig. 1
A comet moving in ether, although its nucleus
is small, drags along a relatively large coma.

Chapter 19 "Comets and Ether", contends that the coma is not gas, ice particles or ions, but rather an atmosphere of ether that the nucleus drags which has become luminous through friction, as it moves through the ether that exists in outer space.

As a comet's nucleus carries a large coma, the earth also carries a large coma (called the "magnetosphere"), with the difference being that, unlike comets, the earth moves so slowly that its coma is invisible, as its motion in ether does not create sufficient friction to produce visible light.

You may have seen on television a video tape of a meteor in the sky of Pittsburgh.[78] The meteor appears much

---

[78] - David Tylor, York Film Ltd., (1993)

like a comet with a very large, bright halo around its head and a long tail. After finding the meteorite, which fell on a car in New York, it was noticed that its size is only a few centimeters in diameter, clearly demonstrating that moving meteors also drag ether as do the earth and comets.

3 - The discovery of the bow shock in front of the earth's magnetosphere at a distance of several thousand kilometers from the earth proves that the earth drags ether (Fig. 2). The bow shock is created by the ether that the earth drags, against the ether in outer space.

Furthermore, scientists have also discovered that the solar wing blowing at a speed of 450 kilometers per second, upon reaching the bow shock is suddenly slowed down and diverted to the sides by a magnetic force. It is as if the bow shock acts as a shield that protects the earth and its atmosphere from the direct impact of the solar wind.

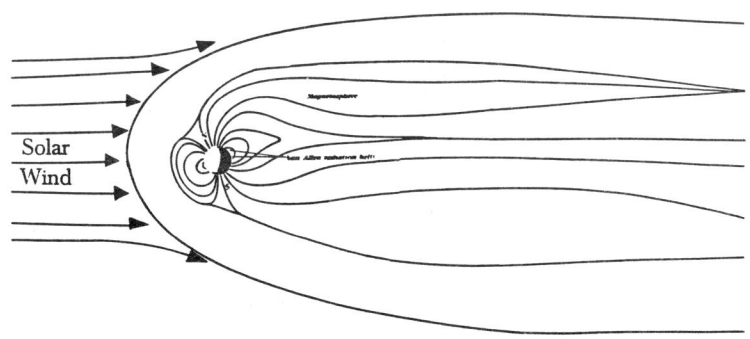

Fig. 2

Fig. 2 shows the bow shock where the solar wind is suddenly slowed down and diverted to the sides by a magnetic force which acts like a wind shield, preventing the solar wind with its normal speed of about 450 kilometers per second from penetrating.

This indicates that the ether near the earth must then move with the earth without being disturbed by such winds as the solar wind or the wind that is created by the earth's motion in ether. Bear in mind that the solar wind is essentially the wind of ether created by the spin of the sun in ether.[79]

4 - Scientists all agree that the earth drags its magnetic field. This means that the ether near the earth must be dragged by the earth.

5 - The experiment carried out by Albert A. Michelson and Edward William Morley provides evidence that the earth drags ether (this will be discussed later on). We know that the earth drags the air atmosphere. We also know that within the air atmosphere the speed or direction of the earth has no effect on the velocity of sound. Similarly, since the earth drags an atmosphere of ether, within the ether atmosphere the speed or direction of the motion of the earth has no effect on the velocity of light. This is exactly what Michelson's experiment shows. His trials, which were carried out at different elevations, all within the air atmosphere, indicate that the direction or velocity of the earth has no effect on the velocity of light. Here we see further evidence that the earth drags ether, because within the air atmosphere all the ether moves with the earth, just as inside an airplane all the air moves with the plane. The speed of sound is unaffected by the speed or direction of the plane.

6 - The phenomenon of stellar aberration is a very clear and convincing indication that the earth drags ether. In the next chapter, we will find out why.

---

[79] - For further details, see Chapter 25, "The Source of the Solar Wind".

Chapter 29

Stellar Aberration

This chapter will show that the phenomenon of stellar aberration is clear evidence that the earth drags ether. Moreover, it will be proven that the explanation for stellar aberration given by relativity supporters is incorrect and that the correct explanation is based on the fact that the earth drags ether. Unfortunately, relativity supporters misrepresent stellar aberration as one of their proofs against the existence of ether, claiming that stellar aberration proves there is no ether and that the earth could not drag ether. Einstein himself claimed that stellar aberration is a proof of his relativity theory,[80] which negates the existence of ether.

James Bradley, in 1725, discovered that the apparent position of the stars is not the same as their actual position. He noticed that a zenith star[81] during a one-year period moved in a circle. The extreme range in a six-month period showed a distance of about 40 seconds of arc. He reasoned that the orbital velocity of the earth has something to do with the deflection. He then concluded that the speed of light and the earth's velocity both cause the deflection.

In 1845, George Stoke, trying to explain the reason for the phenomenon of aberration, came up with the understanding that the earth must at least partially drag ether along its path. Later on, Max Planck went further and concluded that the earth must carry along an ether atmosphere, in much the same way that it carries along an

---

[80] - Albert Einstein, *Relativity, the Special & the General Theory*, Translated by R. W. Lawson, 3rd ed, Methuen & Co. Ltd. London, (1920), p. 49.

[81] The position of a zenith star is not affected by refraction due to the earth's atmosphere.

air atmosphere. He suggested that this could account for stellar aberration.

Today, relativity supporters explain stellar aberration without ether. They advocate the 16th century idea that the deflection of a light ray is simply due to the velocity of the earth and the velocity of light. One of the frequent examples cited is this: on a rainy day a man with an umbrella, standing still, will not get his coat wet; but if he runs, the lower section of his coat will get wet as the direction of the rain drops will be deflected by the speed of the man. The faster the man runs, the more inclined is the position in which he must hold his umbrella to shield himself from getting wet.

The following is how Roemer, in the year 1677, explained the phenomenon.

Let CA denote a ray of light falling on the line BA (Fig. 1).

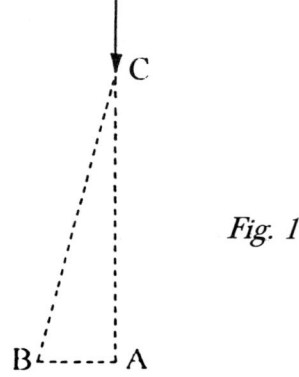

Fig. 1

Suppose also that the eye of the observer is travelling along BA with a velocity which is to the velocity of light as BA is to CA. Then the corpuscles of light by which the object is discernible to the eye at A would have been at C when the eye was at B. The tube of the telescope must, therefore, be pointed in the direction of BC in order to receive the rays from an object whose light is really propagated in the direction of CA. The angle BCA measures the difference between the real and apparent position of the object[82]

Let us look at the explanation more carefully. Fig. 2 shows a ray of light falling through point *C,* the centre at the top of the tube, and the space is assumed to be free of ether.

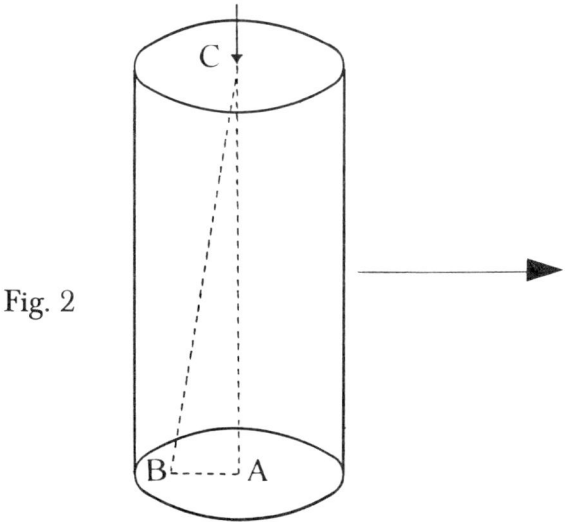

Fig. 2

---

[82] - Phil. Trans. XXXV, (1728), p. 637. Roemer, in a letter to Huygens, 30th Dec. 1677. *C. f.* Correspondence de Huygens, VIII, p. 53.

If the earth were at rest, the light would have fallen at A, but the earth moves forward, so the light falls at B. The time the light ray takes to travel the length of the tube CA is:

$$t = \frac{CA}{c}$$

c is the velocity of light

During the time (t), the earth moves the distance AB.

$$t = \frac{AB}{v}$$

v is the earth's velocity

Thus we have, $t = \dfrac{AB}{v} = \dfrac{CA}{c}$ or

$$\frac{AB}{CA} = \frac{v}{c} = \frac{(30 \text{ km/sec.})}{300{,}000 \text{ km/sec}} = \tan \alpha$$

This angle semi-annually becomes double or 40 seconds of arc.

The result is the same as the angle Bradley found.

After considerable thought given to the above reasoning, it was found that the above explanations could not be true and that the only correct explanation must be based on the earth's drag of ether. Here are the reasons. Let us take a case where a ray of light falls on the ether that the earth drags.

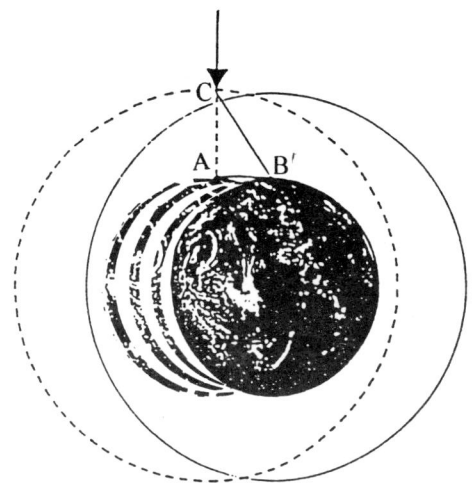

Fig. 3

Suppose a light ray from a zenith star enters the ether around the earth at point C (Fig. 3), in the direction perpendicular to the earth's orbital motion. If the earth were at rest, the light ray would fall at A, on the earth's surface. However, the earth, with the ether around it, is moving forward and carrying the light wave along. By the time the light ray reaches the earth's surface, it has been carried by ether a distance from A to B'. If t is the time that light takes to travel from C to A, then we have:

$$t = \frac{CA}{c}$$

where c is the velocity of light. During the time t the earth travels a distance AB', therefore:

$$t = \frac{AB'}{v}$$

where v is the orbital velocity of the earth. Thus:

$$t = \frac{CA}{c} = \frac{AB'}{v}$$

Therefore we have:

$$\frac{AB'}{CA} = \frac{v}{c} = \tan. a = .0001$$

$$a = 20 \text{ sec.}$$

The angle of deflection is equal to 20 seconds. This angle semi-annually becomes 40 seconds of arc. This result is also in complete agreement with Bradley's observations. The earth's drag of ether also provides a very simple and logical explanation for the deflection of light.

How can we find out which of the above reasons is the correct one? Notice that in Fig. 2 and 3, the earth's motion is towards the right, but the angles of deflection are not on the same side. In Fig. 3, where the earth drags ether, the angle of deflection is to the right, which is the same direction as the earth's motion, whereas in Fig. 2, it is to the left, in the opposite direction to the earth's motion. One deflected to the right and one deflected to the left, making a difference of 40 seconds of arc.

Let us see what the problem is with the explanation given by relativity supporters. It is a known fact that the velocity of light in water, glass or air is different from that in a vacuum. It is also a known fact that if a ray of light enters a moving column of water, the ray is deflected. This was proven experimentally by Fizeau in 1859. He showed that a ray of light is carried or deflected by the motion of the water.

Let us fill the tube in the first case with water (Fig. 4).

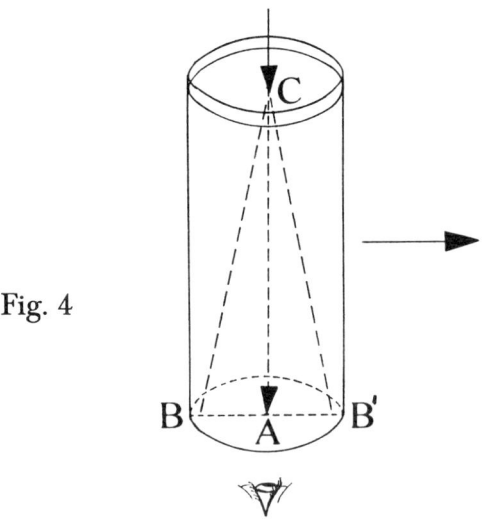

Fig. 4

Since the tube moves to the right with a velocity of 30 kilometers per second, by the time the light ray reaches the bottom of the tube, the tube has moved a distance of *AB'*, and since the tube is filled with water, the light ray is carried a distance equal to *AB'*. This means that if the first explanation is true, then by adding water to the tube, the deflection would change to exactly the opposite direction. A difference of 40 seconds should occur. Sir George Airy in 1871 carried out an experiment in which he filled a telescope with water and found no change in aberration. This shows that deflection has already occurred by the medium of ether that moves with the earth. This also shows that the explanation given by relativity supporters is incorrect. If it were true, then the presence or absence of water in the telescope would make a difference to the apparent position of stars. Furthermore, their explanation not only has not taken into account the fact that the earth drags ether, but also the fact that light rays, before reaching the bottom of the telescope, first travel through hundreds of kilometers of air atmosphere, moving with a velocity of 30 kilometers per second with the earth, and then through the lens of the tube which would, if not fully, at least partially, drag the light rays in the opposite direction. As a result, we should have a deflection of less than 20 seconds of arc. Here we see that the explanation without ether could not possibly be true, and that therefore, stellar aberration is a very clear proof of the earth's drag of ether.

Einstein claimed that stellar aberration supports his theory of relativity. He states that stellar aberration proves that motion affects the length and mass of a body as well as the time of a clock, and that these effects were due to the speed of light. Without explaining how, he made a false claim that aberration supports his theory of relativity, that the mass and length of a moving body and the time of a clock all depend on the speed of light.

Chapter 30

The Michelson and Morley Experiment[83]

A general study of the history of ether reveals thatEinstein used Michelson and Morley's experiment as a basis for his theory of relativity. Einstein claimed that this experiment disproved the existence of ether. After Einstein's claim, in almost all scientific publications, this experiment is represented as an evidence against ether. Here we are going to study the basic ideas of the experiment, to understand why it was carried out and what its findings were, in order to see if it proves or disproves the existence of ether.

A very careful investigation reveals some surprising discoveries, showing that what we read in Einstein's papers and books is misleading. He claimed that the experimental results disproved the existence of ether. On the contrary, it was found that Michelson and Morley's experiments (and many other similar experiments carried out by different scientists), actually confirm the fact that the earth drags ether and that the "negative results"[84] mentioned in Einstein's writings are misrepresentations and distortions of facts.

In 1877, all physicists were very much convinced of the existence of ether. Experiment after experiment successfully led to important discoveries about its qualities. Before Michelson carried out his experiment, George Stoke, a physicist, in 1845 put forth a theory about the reason for aberration, in which he predicted ether at the earth's surface

---

[83] - *Phil. Mag.*, Ser. 5, Vol. 24, No. 151. Dec. 1887, p. 449-463.

[84] - Albert Einstein, *Relativity, the Special & the General Theory*, Translated by R. W. Lawson, 3rd ed, Methuen & Co. Ltd. London, (1920), p. 53. See also Einstein, *The Meaning of Relativity*, Princeton Uinversity Press, (1922), p.29.

to be at rest in relation to the earth.[85] In other words, he predicted that ether in the immediate vicinity of the earth's surface must be dragged by the earth. Hence, at the surface of the earth, no current of ether should be detected. On the other hand, there were two other suggestions, one of them by Fresnel.[86] He suggested that the earth was perfectly permeable to ether, and that ether passes through the mass of the earth, entering one side and emerging from the other. This, again, would produce no current. Finally, there was another suggestion made by Maxwell and later by Michelson himself.[87] They both assumed that the earth cuts through ether, and that, therefore, the motion of the earth in ether should create an ether wind on the earth's surface. Maxwell suggested that the wind must influence the time required by light to travel to and from two fixed points on the earth's surface.

Michelson based his experiment on this idea. He reasoned that the earth moves with a speed of 30 kilometers per second around the sun. If the earth cuts through stationary ether, then the velocity of the wind on the earth's surface should be also 30 kilometers per second. Such a wind should, of course, change the speed of light moving in the direction of the wind or against the wind. For example, if light is transmitted in the direction of the wind, the light will have a higher speed than in any other direction, and if light is transmitted against the wind, the light will have a lower speed than in any other direction. Michelson invented a

---

[85] - A. A. Michelson and E. M. Morley, *Amer. Journ. Sci.*, 34: 333 (1887) or *Phil. Mag.*, (1887) Vol. 24, p. 458.

[86] - See the theory of aberration of light by Fresnel or *Versl. K. Akad. W. Amsterdam.* 1, 74, 1892 by Lorentz.

[87] - This is from a secondary source from Lorentz, "The Relative Motion of the Earth and the Ether", *Versl. K. Akad. W. Amsterdam.* 1, 74, 1892.

device which could detect and even measure the change in the speed of light moving in different directions, which could, in turn, indicate whether or not there is any ether wind. He, together with Morley, carried out experiments which showed that there is no wind with a velocity of 30 kilometers per second. The velocity of the wind they measured was less than a few kilometers per second. Michelson concluded that the experiment proved the earth's drag of ether, and supported George Stoke's prediction.[88]

It must be born in mind that at no time did Michelson doubt the existence of ether. First of all, the reason for carrying out the experiment was not to prove or disprove the existence of ether, but rather, to see whether or not the earth drags ether. Secondly, any negative or positive results of the experiment could only confirm one of the above predictions. For these reasons, the experiment could neither prove nor disprove the existence of ether. However, Einstein misrepresented the facts and claimed that the negative result disproved ether.

After Michelson published the results of his experiments, other scientists with better and more accurate devices repeated the experiments at various heights such as on a balloon or a high mountain, thinking that these heights would be beyond the limits at which ether would be carried by the earth, and that beyond these limits they might find the wind. All their experiments failed to find the 30 kilometers per second wind, indicating that there is no motion of the earth in relation to its immediate envelope of ether.

---

[88] - *Amer. Journ. Sci.* xxii (1881), p. 386; Am. J. Sci. (3), xxii, p. 120. and *Phil. Mag.* S. 5, Vol. 24, No. 151, Dec. 1887.

Henrich Hertz, in 1890, suggested that all ponderable bodies completely drag ether.[89] Max Planck, in 1899, suggested that the earth carries along an ether atmosphere in the same way that it carries along its air atmosphere. Since the air atmosphere extends to least 300,000 feet above the earth's surface, the ether must extend to at least that height as well. The height of a mountain, or the height a balloon could rise above the earth's surface, is only a few thousand feet, which is too low an elevation to be able to detect any motion of the earth relative to the stationary ether.

Today the situation is very different. Scientists have been able to go far beyond the air atmosphere into deep space, and at several million feet above the earth's surface, they have indeed discovered the existence of a wind which they call the "solar wind". They have not related it to ether. Furthermore, the discovery of the magnetosphere, or bow shock, which shields the earth from the direct impact of the solar wind at several million feet above the ground, again indicates that the earth drags ether. The bow shock is created by ether moving with the earth against the solar wind (see Fig. 1). The discovery of the magnetosphere supports Stoke's, Hertz' and Planck's theory that the earth drags ether. It also explains why no motion of the earth relative to ether could exist at such low elevations as the height of a mountain, for, as we have learned, ether travels with the earth up to a height of several million feet above the ground.

---

[89] - See Hertz's 1890 paper. See *Dictionary of Scientific Biography*, Princeton University, vol. VIII, Charles Scribner's Sons, (1973), New York, p. 493.

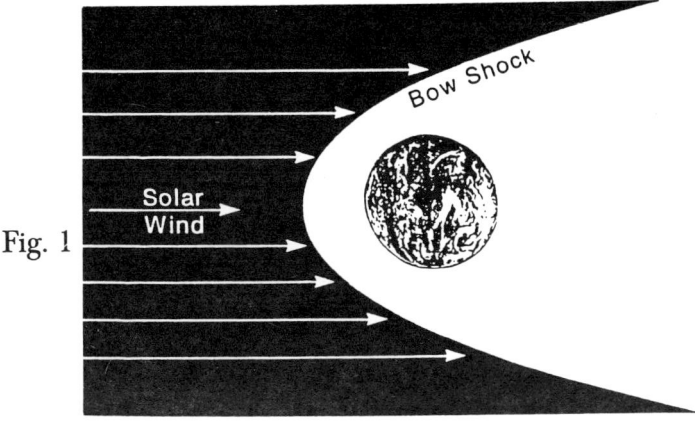

Fig. 1

The discovery of the magnetosphere explains why experiments failed to find the ether wind. It also illustrates that the wind Michelson was looking for exists; however, it is beyond the bow shock and not inside it.

Long before the discovery of the bow shock and the solar wind, Einstein in 1905 claimed that since Michelson's experiment had failed to find the wind, this was proof that ether does not exist. Relativity supporters even up to now, without mentioning the fact that the experiment was carried out at insufficiently low elevations, present it as proof of ether's non-existence. All scientific publications present Michelson's readings as proof against the existence of ether, always emphasising their accuracy but never mentioning the fact that they were carried out at insufficient elevations

Today, relativity supporters explain that the earth's magnetic field shields the earth from the direct impact of the solar wind at a height of 10,000 kilometers from the ground. They claim that the solar wind, moving with a speed of 500 kilometers per second, is checked and then diverted to the sides by the magnetic field of the earth. If the fast-moving solar wind is diverted at such a large distance from the ground, then the slow (30 kilometers per second) wind from the earth's motion through ether must also be checked and pushed to the sides at that distances. How could relativity supporters claim that the results of experiments carried out at low elevations such as that of a balloon or a mountain are proof against the existence of ether?

It is interesting to note that Michelson, from many readings that he made, concluded that:

> *The relative velocity of the earth and the ether is probably less than one sixth the earth's orbital velocity, and certainly less than one fourth.*[90]

It must be emphasised that even the slightest detected ether wind would disprove the foundation of Einstein's relativity. In 1904, Morley and Miller did a new measurement, with an improved interferometer.[91] The experiment was performed in an underground cellar in Cleveland, Ohio. It showed a slow wind with a velocity of less than 3 kilometres per second. In 1905 and 1906, Morley and Miller repeated the experiment,[92] but this time at an elevation of 300 feet above Lake Erie. This experiment confirmed the existence of the ether wind with an even higher velocity. In 1921 and again in 1925, Miller carried out the experiment[93] at Mt. Wilson about 1.8 kilometers above sea level. Miller, as a result of several thousand readings, arrived at the conclusion that the

---

[90] - *Amer. Journ. Sci.* (3), xxii, p. 120, Amer. Journ. Sci. xxii (1881), p. 386., *Phil. Mag.* S. 5, Vol. 24, No. 151, Dec. 1887.

[91] - E. W. Morley and D. C. Miller, *Phil. Mag.*, 9:680 (1905).

[92] - *Encyclopedia Britannica*, 1968, vol.15, p. 368.

[93] - Ibid.

speed of the ether wind was determined to be about 9 kilometers per second. An important question to ask is: why then, since the earth drags ether and hence no relative motion of earth with the surrounding ether should be expected, did the experiments show the existence of a very slow wind (2 to 9 kilometers per second)?

The reason for the wind is the fact that the earth has a magnetic field in which there is a continuous current of ether from one pole to another. The wind that is found in the various experiments is actually the current of ether due to the magnetic activity of the earth. The speed of such a magnetic wind must be in the same order as the reading in Michelson and Morley's experiments, that is to say, a speed that is a small fraction of the orbital velocity of the earth.

The stronger wind that was measured on Mt. Wilson 1.8 kilometers above the sea by Miller could have several causes, including magnetic storms caused by solar flares. For example, the wind of air changes its speed at different times of day as the weather changes and as the sun's position affects the atmosphere. Similarly, we should also expect some change in the speed of the etheric wind at different times for various reasons.

When we consider the facts from this perspective, we see that the results of Michelson and Morley's experiment could not be more positive. Not only did the experiment once again suggest that the earth drags ether, but it also registered the existence of the ether wind from one pole to another, which is called the magnetic field of the earth. However, today one of the frequent reasons used against ether is Michelson's experiment, as if it explained Einstein's theory of relativity and disproved ether. Einstein himself claimed that Michelson's experiment is a proof of his relativity theory. Here is what Einstein wrote:

> *But all experiments have shown that electro-magnetic and optical phenomena, relatively to the earth as the body of reference, are not influenced by the translational velocity of the earth. The most important of these experiments are those of Michelson and Morley, which I shall assume are known. The validity of the principle of special relativity can therefore hardly be doubted.*[94]

Many scientists came to believe that Einstein's claim is a fraud. For example; Dr. L. Essen,[95] who spent a lifetime working on the measurement of time and frequency, writes about Einstein's claim:

> *Insofar as the theory is thought to explain the result of the Michelson-Morley experiment I am inclined to agree with Soddy that it is a swindle; and I do not think Rutherford would have regarded it as a joke had he realised how it would retard the rational development of science.*[96]

---

[94] - Albert Einstein, *The Meaning of Relativity*, Princeton University Press, (1923), Princeton, New Jersey, p. 29.

[95] - Dr. Essen of the NPL, built the first caesium clock in 1955 and determined the velocity of light by cavity resonator.

[96] - Louis Essen, "Relativity - Joke or Swindle", *Electronics and Wireless World*, v. 94 (Feb. 1988), p. 126-127.

Chapter 31

Einstein's Rejection of Ether and Evidence of Fraud

So far, we have studied the phenomena which support the existence of ether. In the following few chapters, we will investigate reasons used against ether in order to find out why scientists abandoned the concept. After some research, it was found that it was Albert Einstein who lead scientists to abandon ether. Here is an example of what Einstein wrote: *"This is the moment to forget the ether completely and try never to mention its name."* [97]

All physicists agree that it was Albert Einstein[98] who published a series of articles eventually leading scientists to change their understanding of ether.[99] For example, P. A. M. Dirac writes: *"Einstein destroyed the ether. . ."*.[100] Collier Encyclopaedia writes:

> *Einstein's theory of relativity appeared in 1905, and considering its radical nature, won general acceptance in a surprisingly short time.... This came about in part because the theory of relativity showed....that the ether theory had almost certainly to be abandoned.*[101]

---

[97] - Albert Einstein and Leopold Infeld, *The Evolution of Physics*, Simon and Schuster, New York, (1938), p.184-185.

[98] - Sir Edmund Whittaker, *A History of the Theories of Aether and Electricity*, New York, (1962), p. 8.

[99] - As an example, see *Encyclopedia Britannica*, 1967, v. 8, p. 747, which states that it was Einstein who led scientists to abandon ether.

[100] - *Van Nostrand's Scientific Encyclopedia*, sixth edition, New York, Van Nostran Reinhold Company, p. 2428.

[101] - *Collier's Encyclopedia*, (1987), P. F. Collier Inc., London, New York, v. 14, p. 627.

When I studied all of Einstein's relativity papers including the one he wrote in 1905, I did not find even a shred of justified scientific evidence presented against the existence of ether. Later, while investigating the reason why ether was abandoned, I became aware of something highly unusual in the source books. There is a lot more to the ether story than what meets the eye by just studying text books or famous scientific publications.

The following is a summary of some astonishing and bitter discoveries. What you are about to read about Einstein's works is quite different from what you have been taught in school or universities. I was not happy to make these discoveries, for I had always regarded Einstein as the greatest scientist mankind has ever produced. From my childhood, I had been taught to love and admire his achievements. During my research, however, I slowly realised that what I had been taught is different from reality. These revelations changed my view of Einstein. Much of the evidence affecting my view of Einstein will be presented in this chapter.

In my University years, we were taught that the case against ether is closed, on the grounds that scientists have disproved its existence. However, my research reveals another picture. I discovered many controversies concerning the alleged "proofs" against ether. I learned that at the time, the vast majority of scientists were strongly convinced of its existence. Moreover, I realised that, despite the disagreement of scientists, Einstein and a few of his supporters were able to create legendary fame and popularity for themselves by publicising ideas such as time travel, the idea that man may one day be able to travel into the future and the past. This appealed to the imagination of the public very much and was given immense publicity. Thus through the power of the media, Einstein and few of his friends were able to dictate their ideas as standards of truth. One such idea is that ether does not exist.

My point is that not only does ether exist, but scientists before Einstein knew this and were using ether to explain the behaviour of light, electricity and magnetism. With Einstein's denial, science was arrested in these areas and progress was halted by the ensuing confusion. We followed Einstein down preposterous paths and became lost. I consider this blind following of a man whom we erroneously elevated to the status of an icon as nothing short of a grievous loss.

My investigations led me to some other astounding discoveries about Einstein. I discovered that the relativity theory did not originate with him. This was a great surprise to me; who indeed does not link relativity with the name of Albert Einstein? Further investigations disclosed that all the other famous works attributed to Einstein, including the relativity theory, the photon theory, and the well-known formula of $E = mc^2$, actually belong to other scientists and were not Einstein's. Furthermore, in Einstein's papers I found unrelated conclusions, mathematical misapplications and, above all, many misrepresentations. I noticed also that, in order to disprove evidence of ether, Einstein methodically fabricated false information against it. I realised that before Einstein, there had been absolutely no problem about the existence of ether. Einstein misled scientists into abandoning ether and replacing it with idea of the empty space. The relativity theory which had been founded on the basis of ether, Einstein used as evidence against ether. All the relativity formulas originally developed by Lorenz on the basis of ether, Einstein misinterpreted as evidence against ether. Einstein sacrificed and discarded many fundamental laws and principles of physics in order to accommodate his own ideas. Years of research and investigation into Einstein's work has led me to conclude that it amounts to the greatest fraud of this century.

I have discovered that many scientists have disproved Einstein's ideas but they are not mentioned in any text books. Some have bitterly complained about the suppression of facts by the press and famous scientific publications, all of whom speak in favour of Einstein. Some of these scientists have asked for reasons for this suppression. For decades, it was widely publicised that only three persons have understood Einstein's papers. For these papers that they did not understand the media praised Einstein as the greatest scientist humanity has produced and he was repeatedly awarded Nobel prize. After examining the evidence very carefully, I conclude that the abandonment of ether was not based on any true scientific reasons, but rather on a series of misrepresentations and mischievous manipulations by vested interest.

In this chapter, evidence for this conclusion is presented in condensed form in order for the reader to see the whole picture comprehensively. Footnotes provide all necessary references to the original texts. The appendices offer more information and discussion on the subject.

Facts about Einstein and his works

The following evidence can be easily verified. Most of it is already known as isolated cases, but when assembled altogether like pieces of a puzzle, it presents a very clear picture of the true reality of Einstein's works. Since it was not possible to include all the evidence in this book for reasons of space, only the most obvious and accessible examples are presented here.

A few years prior to publishing his relativity paper, Einstein was known to be a slow student at the Swiss Federal Institute of Technology and had even failed one of the normal entrance examination.[102] He was a student who admitted to passing his final graduation exams by using the notes of his fellow classmate Marcel Grossmann, and was known to break school rules and skip classes.[103]

In 1902, with the help and intervention of Marcel Grossmann's father, he was employed in the Swiss Patent Office. Before being employed in the patent office, in 1902, Einstein[104] wrote two papers which he himself admitted were worthless[105] and were rejected. After being employed in the patent office, he wrote many papers. During the year 1905, he wrote twenty-one review papers plus his famous papers and a Doctoral thesis[106] which no one could truly understand. In that same year, in a one-month period, he wrote seven papers including his famous paper on relativity. These facts lead us to this question: what happened in those few years to change Einstein into a super scientist of great historical significance?

Einstein, during the period of his life in which he worked in the patent office, wrote papers on various subjects which other scientists had worked on over long periods of time. The controversies which these papers unleashed are

---

[102] - Ronald W. Clark, *Einstein: The Life and Times*, New York, World Publishing Company. (1971), p. 25.

[103] - Ibid 37.

[104] The Swiss Patent Office was founded in 1888. See also fn. 102, p. 45.

[105] - ibid, p. 52.

[106] -"A New Determination of Molecular Dimensions", a thesis which consisted of the treatment of differential equations, being thus of a mathematical nature, but which won him a Doctorate in Physics. See also fn. 102, p. 49.

unparalleled in the history of science. Various scientists tried bitterly and in vain to prove that the ideas and findings published by Einstein actually did not belong to him.

Did all of Einstein's famous works originally belong to him, or did they belong to other scientists?

Evidence shows:
1. the famous formula $E = mc^2$, did not belong to Einstein, but to Poincaré;
2. relativity did not belong to Einstein, but to Lorenz and Poincaré;
3. the idea of the fourth dimension was not Einstein's idea, but Poincaré's, with further contributions by Minkowski;
4. the idea of going back in time or going into the future was not Einstein's idea but, Minkowski's; the chain reaction was not Einstein's discovery;
5. the atomic bomb was not Einstein's discovery;
6. Einstein's letter to President Roosevelt did not cause the production of the atomic bomb;
7. the photon theory did not belong to Einstein;
8. the photoelectric effect formula did not belong to Einstein;
9. the predictions which Einstein made concerning light did not originate with him;
10. the principle of equivalence was not Einstein's idea but Max Plank's;
11. Einstein's hypothesis originally did not belong to him; and
12. Einstein's two famous papers on the statistical theory of heat originally did not belong to him, but to Willard Gibbs.

Every one of these facts will be dealt with separately as follows. It must be born in mind that the above-mentioned works include all of Einstein's famous papers.

$E = mc^2$ was not Einstein's formula

This is the famous formula everyone associates with Einstein. Rudakov points out the fact that it was developed by Poincaré and others before Einstein:

> *That the mass increase or the total mass of a particle may be equivalent to energy in accordance with a formula which has the configuration*
>
> $$E = mc^2,$$
>
> *or a very similar configuration containing the square of the velocity of light, was proposed by Poincaré, Hasenohrl and others before Einstein. Langevin suggested the formula independently of Einstein. . . . It is not correct to assert that the mass increase and the mass-energy equivalence have been established by Einstein.*[107]

Science historian, Sir Edmund Whittaker,[108] writes that the formula $E = mc^2$ actually belongs to Poincaré and not Einstein. Poincaré, in 1900, had published that "... *electromagnetic energy might possess mass density equal to $1/C^2$ times the energy density. That is to say*

$$E = mc^2$$

---

[107] - N. Rudakov, *Fiction Stranger than Truth*, Australia, published by N. Rudakov, (1981), p. 161.

[108] - Sir Edmund Whittaker, *Aether and Electricity*, Thomas Nelson and Sons Ltd., (1910), p. 51.

where E equals energy and m equals mass." [109]

On this basis, Poincaré had made some predictions that 8 years later were confirmed by G. N. Lewis.[110]

Einstein, in 1905, five years after Poincaré, mathematically manipulated a hypothetical case, suggesting[111] that:

$$K_0 - K_1 = L\{(1/1 - v_2/c^2) - 1\}$$

a formula which, only after some approximation, becomes similar to Poincaré's formula. In this paper,[112] there is a clear evidence of deception[113] as to render the entire paper worthless from the point of view of science. For example, in the beginning Einstein takes Poincaré's formula as his starting point, moves it slightly off-centre and then works his way back so he produces the original formula. In other words, Einstein introduces an equation which is in reality the same as Poincaré's formula, but disguised in a form which is not easily recognizable. Then after some mathematical manipula-

---

[109] - H. Poincaré, *Archives Neerland*, Sci. Exactes Nat. 2, v (1900), p. 252. See also A. I. Miller, Ref. 1. P. 40-45.

[110] - *Phil. Mag.* xvi (1908), p. 705.

[111] - *Ann. d. Phys.* xviii (1905), p. 639.

[112] - See the Appendix for an English translation of Einstein's paper called "Does the inertia of a body depend upon its energy-content?"

[113] - See Rudakov, *Fiction Stranger Than Truth*, fn. 104, p. 161-175.

tion[114] he produces a formula which could easily be converted into Poincaré's formula.

Professor Herbert E. Ives points out the fact that Einstein "*introduces the very relation that his derivation was supposed to yield.*" Ives proves that Einstein's paper is defective:[115]

> *The mass equivalent of radiation is implicit in Poincare's formula for the momentum of radiation, published in 1900, and was used by Poincare in illustrating the application of his analysis. The equality of the mass equivalent of radiation to the mass lost by a radiating body is derivable from Poincare's momentum of radiation (1900) and his principle of relativity (1904). The reasoning in Einstein's derivation, questioned by Planck, is defective. He did not derive the mass-energy relation.*[116]

---

[114] - Otto Luther, *Relativity is Dead*, California, Key Research Corp., (1986), p. 129.

[115] - H. E. Ives, *Jour. Opt. Soc. Amer.* xlii (1952), p. 540.

[116] - Herbert E. Ives, "Derivation of the Mass-Energy Relation", *J. Opt. Soc. Am.*, v. 42, No. 8, Aug. 1952, p. 540.

After reviewing in detail Einstein's paper, Ives points out that the work is nothing more than a manipulation of mathematics.

The realisation that Einstein's mathematical steps are circular confirm the author's understanding that all of Einstein's papers, including his relativity paper, more or less contain this problem. This explains how Einstein was able to write papers which, on the surface, appear to have been written independently, but, in fact, were set-ups to reach conclusions which had been arrived at by other scientists.

Relativity Theory does not belong to Einstein

After decades of controversy over to whom the relativity theory originally belonged, scientists generally agree that it belongs jointly to Poincaré and Lorenz. To cite a few examples, the reputable science historian, Sir Edmund Whittaker, writes:

> *Einstein published a paper which set forth the relativity theory of Poincaré and Lorenz with some amplification and which attracted much attention.*
> 117

As another example, Elie Zahar, in his book "Einstein's Revolution" writes:

---

[117] - See Whittaker, *A History of Theories of Aether and Electricity,* New York, (1962), p. 40.

> In what follows I propose to defend a thesis as brutally simple as Wittaker's, namely that Poincaré did discover special relativity, . . .[118]

Lord Kelvin, one of the greatest scientists of the time, writing on Lorenz' and Poincaré's relativity theory, states:

> Now, in comparing this theory with that of Einstein, I want to say emphatically that the physical contents of the two - as I understand the notion of "physical content" - are identical.[119]

Before Einstein's paper, Lorenz had already published many articles about relativity and had developed the formula for relative length, mass[120], etc. The relativity formulas in use today were originally developed by Lorenz, and not by Einstein.[121] All the mathematical formulas for relativity in Einstein's paper were the same as those in Lorenz's and Poincaré's papers. Collier's Encyclopaedia, says:

> To some extent the independent work of Lorenz and Poincaré preceded that of Einstein, although it was not known to him; much of the mathematical content of the theory was first published by them. . .[122]

---

[118] - Elie Zahar, *Einstein's Revolution: A Study in Heuristic*, Open Court Publishing Company, La Salle, Illinois, (1989), p. 150.

[119] - *Kelvin's Baltimore Lectures and Modern Theoretical Physics*, edited by Robert Kargon and Peter Achinstein, The MIT Press, Cambridge, London, (1987), p. 386.

[120] - *Amst. Proc.* vi (1904), p. 809.

[121] - See Sir Edmund Whittaker, *A History of the Theories of Aether and Electricity*, New York, (1962), p. 30.

[122] - *Collier's Encyclopedia*, (1990), vol. 19, p. 712.

Or, the McGraw-Hill Encyclopaedia of Science and Technology writes:

> ... *the mathematical structure of the Lorenz-Poincaré is the same as that of the theory advanced by Einstein in 1905, independently of Lorenz's work of 1904, and of Poincaré's work* ... [123]

Herbert Dingle, Professor Emeritus of History and Philosophy of Science at the University of London writes:

> *..from 1904 until 1919 the 'relativity theory' was ascribed to Lorenz, not to Einstein. . . . In 1919. . . overnight 'the relativity theory of Lorenz' became 'Einstein's special relativity theory'.*[124]

Why, then, did Einstein gain fame and credit for a work that originally belonged to Lorenz and Poincaré? It is interesting to note that all of the relativity formulas used today as Einstein's formulas originally belonged to Lorenz. It is interesting to note also that after reading Einstein's relativity paper, I realised that it is full of misrepresentations which lead me to question whether the paper is shear fraudulence. The evidence will be discussed in detail later on in this book.

The following is a presentation speech by professor S. Arrhenius, the Chairman of the Nobel Prize Committee

---

[123] - The fact has already been established that the mathematical aspect of Einstein's paper is almost a duplicate of Lorenz' and Poincaré's papers. As an example, see: *McGraw-Hill Encyclopedia of Science and Technology*, (1992), v. 15, p. 284.

[124] - Herbert Dingle, *Science at the Crossroads*, Martin Brian & O'Keefe, London, (1972), p. 95.

for Physics in 1921. The speech may give you some ideas how Einstein, by the abuse of the media, became famous.

Your Majesty, Your Royal Highness, Ladies and Gentlemen.

There is probably no physicist living today whose name has become so widely known as that of Albert Einstein. Most discussion centres on his theory of relativity......

The presence here of hundreds of people to greet a distinguished man of science is part of something without parallel in our American life. When newspapers everywhere continue, day after day, to give front page space to a man whose work does not directly touch the lives of people, it signifies something unusual. The public itself would first catch the humor of a suggestion that it knows anything about relativity; and yet the warm interest in the man who has given us relativity continues.[125]

The Fourth Dimension and Time Travel are not Einstein's Ideas

Today the general public holds a misconception that the fanciful aspect of the theory of relativity (which involves the fourth dimension, or the notion that a time machine may one day make it possible to go into the future or the past),

---

[125] - *Albert Einstein's Theory of General Relativity,* Edited by Gerald E. Tauber, Crown Publishers, INC, New York, (1979), p. 35.

belongs to Einstein. However, after some investigation, I found that neither idea can be attributed to him.

After Einstein published his relativity paper, many scientists realised that it misrepresents and distorts the facts. They could do nothing about it, however, because Herman Minkowski, Einstein's friend and mathematics teacher, was able to provide an interpretation which saved Einstein and turned everything into his favour by introducing the idea of the fourth dimension. Here is what Einstein himself admitted:

...the form given to the special relativity theory by Minkowski, which mathematician [sic] was the first to recognise clearly the formal equivalence of the space-like and time-like co-ordinates, and who made use of it in building up the theory.[126]

The fourth dimension idea then led to the fiction of going into the future and the past, which captured the imagination of the public at large.

It is interesting to note that the idea of four-dimensional space originally belonged to Henry Poincaré, who came up with it in 1905, while formulating his hypothesis about gravitation.[127] Later, Minkowski[128] changed Poincaré's four-dimensional space co-ordinates into four-dimensional

---

[126] - Einstein, "Principle of Relativity", Translated by M.N. Saha & S. N.. Bose,University of Calcutta, (1920), p. 89. From *Annalen der Pysik* 4.49. 1916.

[127] - Henry Poincaré, "Sur la Dynamique de l'Electron", *Comptes Rendus de l'Academie des Sciences*, t. 140, p. 542, (5 Juin 1905); or C. M. Kilmister, *Special Theory of Relativity*, Oxford: Pergamon Press, (1970), p. 176; or *Aeures de Henri Poincaré*, Gauthier- Villars, Editeur-Imprimeur-Lbraire, (1954), p. 542.

[128] - Michael White & John Gribbin, *Einstein*, Simon & Schuster Ltd., (1993), p. 39.

time/space co-ordinates. In other words, Minkowski added the idea of space changing into time and time changing into space and invented the four-dimensional time/space continuum.[129]

The Photon Theory does not Belong to Einstein

The photon theory commonly attributed to Einstein is originally Newton's centuries-old "corpuscular theory".

According to Einstein, the energy of light, instead of being distributed through space, is concentrated in small packets, called "photons". This is exactly Newton's "corpuscular theory", with new words, a new label, and Einstein's name as author. Furthermore, in 1804, a century before Einstein, J. Soldner[130] suggested the theory. Walter Ritz (1878-1909) advanced the "ballistic" theory. Finally, Max Planck in 1900, five years before Einstein, also advanced the photon theory.

The existence of the photon was first suggested by Planck's famous research, about 1900.....[131]

In Appendix F, "Photon Theory: Evidence of Misrepresentation", you will find many clear and irrefutable proofs that light is not a photon, but a wave.

---

[129] - *The New Encyclopedia Britannica*, v. 8, p. 166; or *Dictionary of Scientific Biography*, v. IX, p. 414.

[130] J. Soldner, *Berliner Astr. Fahrb.* (1804), p. 161.

[131] - *Van Nostrand's Scientific Encyclopedia*, 6th ed., by Considine editor, (1983), New York, p. 2231.

The following is all that the Encyclopaedia Britannica writes under "photon":

> *photon: Albert Einstein's name for the theoretical quantum (discrete unit) of radiant energy. Max Karl Planck had posited that light (q.v.), for example, is emitted in quanta rather than continuously.*[132]

In this case, as in previous cases, Einstein receives credit for a theory that did not originate with him. Later on you will see clearly how Einstein was able to take credit for other papers that originally did not belong to him.

Were Einstein's papers[133] on the statistical-kinetic theory of heat copies of Willard Gibbs papers?[134]

It is a curious fact that Einstein, in six months, during the time he worked in the patent office, wrote four of his most important papers. Three of these (relativity, the photon theory, and mass-energy equivalence), contained results which were not his own, as we have seen. What about the fourth paper? Are its results also derived from other scientists?

My investigation revealed the same pattern of fraud repeated again and again. Here is what science historian, Sir Edmund Whittaker, wrote about these papers:

---

[132] - *Encyclopedia Britannica*, (1968), V. 17, p. 1000.

[133] - *Ann. d. Phys.* (4) xvii (1905), p. 549. and *Ann. d. Phys.* xix, (1906), p. 371.

[134] - *Ann. d. Phys.* ix (1902), p. 417, and *Ann. d. Phys.* xi (1903), p. 170.

. . papers[135] *on the statistical-kinetic theory of heat, in which, however, Einstein had only obtained independently certain results which had been published a year or two earlier by Willard Gibbs.*[136]

We must bear in mind that Einstein, before writing these papers, had never carried out an experiment on the subject.

Is this what Einstein learned while working at the patent office?

Einstein did not discover the Chain Reaction.
Einstein did not invent the Atomic Bomb.

My research indicates that Einstein made absolutely no contribution to the discoveries of either the chain reaction or the atomic bomb, yet he is credited for them in many scientific publications. Here follows the trail of my research.

Firstly, Otto Hahn and Fritz Strassmann, two German chemists who followed the Irene Joliot-Curie findings, concluded that they had split the uranium atom, which at the time was unthinkable to all physicists, including Einstein.[137]

---

[135] - *Ann. d. Phys.* ix (1902), p. 417; xi (1903), 170.

[136] - See Sir Edmund Whittaker, *A History of the Theories of Aether and Electricity*, New York, (1962), p. 8. See also Clark, *Einstein, The Life and Times,* The World Publishing Co., New York, (1971), p. 53.

[137] - See a German scientific weekly, *Die Naturwissenschaften* (The Natural Sciences), December 22, 1938, and William L. Lawrence, *Men and Atoms*, Simon and Schuster, New York (1959), p. 17, 27.

In 1939, Otto Frisch confirmed the findings and, during an experiment, discovered that the energy released is 100 million times as great as the burning of one hydrogen atom in oxygen.[138] Walter Zinn and Leo Szilard, in March 1939, arrived at the idea of the chain reaction and the possibility that,

> ...the vast energy released in the fission of uranium could be utilised either in an atomic power plant or in an atomic bomb equal in destructiveness to thousands of conventional bombs of the same size.[139]

It is interesting to note that during all this period, Einstein was continuously preoccupying himself with his cosmos and unified field theories, isolating himself from the mainstream of physics.[140] Professor Aage Bohr, a close friend of Einstein, writes that he

> ..was deeply involved in his own work and I hardly think that he was following the current development in nuclear physics.[141]

Why has Einstein been credited as if he had a great part in the discovery of the chain reaction?

It must be borne in mind that Otto Frisch, during an experiment, observed that a tremendous amount of electrical

---

[138] - Otto Frisch, *Nature*, Jan. 16, 1939 and Feb. 18, 1939; Otto Frisch and Lise Meitiner, *Nature*, Feb. 11, 1939; *Men and Atoms*, by William L. Lawrence New York, p. 31.

[139] - Ibid p. 38.

[140] - See Clark, *Einstein, The Life and Times*, The World Publishing Co., New York, (1971), p.552.

[141] - Professor Aage Bohr, Author, September 23, 1970.

energy is created by splitting the uranium atom. This discovery was made by experiment and not through the formula $E = mc^2$. Why then, in many major scientific publications, has the discovery of the chain reaction and the atomic bomb been linked to the formula $E = mc^2$ and to Einstein, as if the discovery had been made through the formula and by Einstein? Furthermore, even if there were a relationship between the chain reaction, the atomic bomb and the formula, the credit should go to Poincaré, not Einstein, since Poincaré developed it.

Einstein did not start the Manhattan Project.
Einstein's letter did not start the production of the Atomic Bomb

It has been claimed that in 1939, Leo Szilard wrote two letters, one of which he himself signed and the other he gave to Einstein to sign,[142] on the dotted line, to be presented to President Roosevelt. It has also been claimed that Dr. Alexander Sachs presented both letters to President Roosevelt. The letters described the development of the chain reaction and the possibility that it might lead to the development of an atomic bomb. Many publications claim that Einstein's letter started the Manhattan project which, in turn, produced the atomic bomb. Einstein himself said:

> *My participation in the production of the atomic bomb consists of one single act: I signed a letter to President Roosevelt.*[143]

---

[142] - See Clark, *Einstein, The Life and Times*, The World Publishing Co., New York, (1971), p. 557.

[143] - Ibid, p. 562.

After careful investigation, it soon became clear that in this case, as in all other cases related to Einstein, there is a controversy.[144] On the one hand, many publications claim that Einstein's letter had a very great effect; that it did, in fact, launch the Manhattan Project which produced the atomic bomb. On the other hand, many American and British officials, scientists, historians and key officials working on the project said that the Manhattan project began because of a British report confirming the feasibility of the bomb's production and its effectiveness in winning the war. These people claim that Einstein's letter had no effect, did not create the Manhattan project and that America would have made the bomb with or without Einstein.

For example, the Director of Scientific Research and Development, and later the key person in the project, Vannevar Bush, stated: *"...the show was going before that letter was even written."* [145] Arthur Compton, a key worker in the field, said that the effect of Einstein's letter *"was to retard rather than to advance the development of American uranium research."*[146]

My research also showed that until a few weeks before the starting date of the Manhattan project, the American scientists did not believe the production of an atomic bomb would ever be possible. The government and military hardly gave it any attention. From the military point of view, even if it were ever possible to build the bomb, it would be too heavy to be transported by air. A heavy bomb that could not be transported by air would not be considered a decisive weapon for winning a war.

---

[144] - Ibid, p. 563.

[145] - Vannevar Bush, *Boston Globe*, December 2, 1962, or "Einstein, The Life and Times", (1971), p. 549.

[146] - Ibid, (1971), p.562.

The Manhattan Project started only after the British scientists working on the bomb in Britain discovered that the amount of separated uranium required for a bomb weighed pounds rather than tons. The British committee (The M.A.U.D. Committee) in 1941 concluded that production of the bomb was possible and that its weight could be small enough to be carried by air. The British committee recommended immediate action: *"Indeed, it would be a decisive weapon. Construction should begin without further delay."*[147] For lack of resources and fear of German bombings, the committee recommended that the bomb be built in the U.S. or Canada. When the report of the committee was communicated to Washington, it surprised the U.S. Government, scientists and the military, because it promised not only that production of the bomb was possible, but also that the bomb would be a decisive weapon for winning the war. Furthermore, it gave a concrete outline of how to produce it. President Roosevelt, after several days of consultation, sent a letter to Churchill proposing that the two countries co-operate in building the bomb. In less than a few weeks, the Manhattan Project was begun.

With these facts in mind, it is clear that the claim that Einstein's letter started the Manhattan project is a misrepresentation by the media designed to enhance the image of Einstein. It is interesting to note that at the time Einstein allegedly signed the letter, and even up to a year after the Manhattan Project had begun, he and his friend Bohr[148] believed that it would be impossible to build an actual atomic bomb, or to harness nuclear energy. They

---

[147] - Lansing Lamont, "Day of Trinity" McClellant and Stewart Ltd, New York, 1st edition, (1965), p.

[148] - Bohr-Chadwich quoting Cockcroft, *Biographical Memoirs of Fellows of the Royal Society*, Vol.9, p. 45.

believed that to separate enough of the isotope from chemically identical atoms to make a bomb or a nuclear reactor would take an unforeseeably long period of time, if ever, making it practically impossible for the war effort. Einstein admitted: *"I did not, in fact, foresee that it would be released in my lifetime."*[149] It is also interesting to note that Einstein, until the bomb was built, knew nothing about the project. The information was kept secret from him because the American government and many scientists were very much suspicious of him and did not trust him at all.[150]

The following is a copy of an article from a history book.[151] The article is a typical example showing how facts have been misrepresented. Einstein's name and picture have been presented as if he were the creator of the atomic bomb.

---

[149] - Einstein, "Atomic War or Peace", *Atlantic Monthly*, November 1945.

[150] - See Clark, *Einstein, The Life and Times,* The World Publishing Co., New York, (1971), p.564 - 565.

[151] - *Time-Life Books: History of the Second World War*, Prentice Hall Press, New York (1989), p. 427.

THE CREATION OF THE BOMB

The chain of events that led to the creation of the bomb began with scientific experiments conducted during the 1930s. As physicists in the united States and on the Continent labored to discover how the atomic structure of matter might be transformed, uranium and plutonium seemed the most promising targets.

On December 9, 1939, two scientists in Berlin succeeded in accomplishing something extraordinary: they split the nucleus of an atom. By March 1939, refugee Hungarian physicist Leo Szilard produced a laboratory-scale chain reaction using uranium. This significant advance meant that nuclear power—and an atomic weapon—was indeed theoretically possible.
....
As experiments continued to reveal the mysterious characteristics of uranium and plutonium, a burning moral question persisted: what should the scientists do with their potent new knowledge? Their answer to the question was dictated by concern that an enemy might develop the bomb first.

The scientists, albeit meeting with laboratory success, failed to spark government interest.. They persuaded Albert Einstein, whose pioneering theory outlined the relationship between matter and energy, to sign a letter to President Roosevelt. The letter which described the development of nuclear chain reaction that would generate vast amount of power, and warned of the consequences—was received on October 11, 1939. The eventual result was presidential approval of a project named the Manhattan Engineer District (an obscurity intended to detect attention from its real purpose), which in turn, according to Hungarian scientist Edward Teller, turned "the whole of the United States" into a nuclear factory.

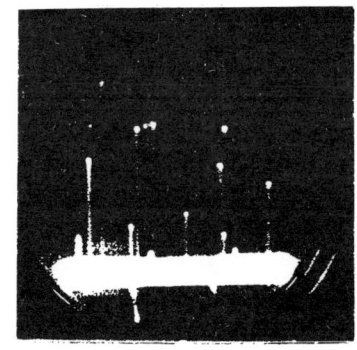

The vertical pulses on a voltage indicator record the bursts of energy released when U-235 (an isotope of natural uranium) atoms are split. By 1940, nuclear scientists possessed the raw material and the crude process for creating the most powerful energy source ever conceived by mankind: U-235 plutonium set loose in an instantaneous chain reaction.

Albert Einstein makes notes in his study at Princeton University in 1943 it was Einstein's letter to president Franklin Roosevelt in 1939 that got the United States started on the Manhattan Project. The action ultimately taken, however, moved the gentle Einstein, years later, to remark: "I made one great mistake in my life, when I signed the letter to President Roosevelt."

Einstein's Star Shift Predictions do not originate with him

In 1805, the Munich astronomer, Professor J.Soldner[152] advanced the idea that light has mass, and that a ray of light from a star will be deflected when it passes by the sun. He even calculated the amount of the shift. Einstein, a century later, published the prediction as if it were his own.[153]

Furthermore, the apparent shifting of stars had already been known and explained by Willebrod Snell, who discovered the law of refraction in 1621. Newton, in 1687, also discussed the star shift.

Therefore, it is very clear that predictions Einstein attributed to himself were not, in fact, his own. He received credit for them through misrepresentation and false pretence.

Einstein's hypothesis concerning the Perihelion of Mercury originally did not belong to him

Einstein published this hypothesis in 1920.[154] However, A. Hall, in 1894, had already introduced the same hypothesis.[155]

---

[152] - *Berliner Astr. Jahrb.* (1804), p. 161.

[153] - Albert Einstein, *Relativity, the Special & the General Theory*, Translated by R. W. Lawson, 3rd ed, Methuen & Co. Ltd. London, (1920), p. 75.

[154] - Ibid, p.124-125.

[155] - A. Hall, "A Suggestion in the Theory of Mercury", *The Astronomical Journal*, No. 319, Vol. XIV, Boston, June 2, 1894, No.7.

The principle of equivalence does not belong to Einstein

The principle of equivalence, which is the basis of Einstein's theory of gravitation (called the "General Theory of Relativity"), was not Einstein's idea, but Max Planck's, and was published six months before Einstein's paper.[156] The principle of equivalence was simply a new label for Planck's idea that "...*all energy has inertial properties...*" and therefore "...*all energy must gravitate.*"[157]

The Photoelectric Effect Formula was not Einstein's Formula

A scientific principle commonly taught in secondary school states that there is never 100% efficiency in the conversion of energy from one form to another; there is always some energy loss. For example, when we throw a ball, the kinetic energy gained by the ball is not all the energy used. Some energy is used to move the hand, some is spent as heat, some as friction or air resistance, etc. Therefore, we can write:

E, total energy used = kinetic energy of the ball + Q, energy loss, or:

---

[156] - *Jahrb. d Radioakt.*, iv (4 Dec. 1907), p. 411 ; A. Einstein, *Ann. d. Phys.* xxxv (1911), p. 898.

[157] - See Sir Edmund Whittaker, *A History of the Theories of Aether and Electricity*, New York, (1962), p. 152.

$$E = \tfrac{1}{2}mv^2 + Q$$

(m is the mass of the ball and v is its velocity)

Since in the photoelectric effect, light energy causes some electrons to escape from a metal surface, then in this case light energy is converted into kinetic energy, and we can write a similar formula:

E, light energy used = kinetic energy of the electron + Q energy loss

In other words, some of the energy of light is converted into the kinetic energy of the electron and the rest is lost by conversion into other forms of energy, or:

$$E = \tfrac{1}{2}mv^2 + Q$$

(m is the mass of the electron and v is its velocity)

Einstein used Planck's formula: $E = hf$, replacing E by $\tfrac{1}{2}mv^2 + Q$, hence presenting Planck's formula in a disguised form which became known as the photoelectric formula:

$$hf = \tfrac{1}{2}mv^2 + Q$$

    Therefore, Einstein's photoelectric formula is in reality Planck's formula that has been glorified and displayed in science textbooks as one of the greatest achievements in science. It is interesting to note that for this formula, Einstein's supporters repeatedly awarded him prizes, including the Nobel Prize.

It must be noted that Einstein attempted to take credit for Planck's formula $E = hf$, but he failed. He pretended to have found the value of h independently. However, in this case he could not deceive scientists, because the constant "h" was already known to all physicists as "Planck's constant".

Interestingly, Einstein claimed that the photoelectric formula is a proof that light is a particle. It will be illustrated later how his claim is a misrepresentation designed to prove his relativity ideas. See Appendix F, 'Photon Theory: Evidence of Misrepresentation'.

So far, what we have learned is that Einstein by misrepresentation was able to take credit for ideas that originally did not belong to him. Further evidence of misrepresentation and fraud will be discussed later on. However, before presenting this evidence, it will first be necessary to know the history of relativity, how it was developed and how Einstein changed it.

Chapter 32

Einstein's Relativity

The following is a brief history of how the theory of relativity was founded on the basis of motion through ether. It will then be discussed how Einstein disguised this theory.

Since space is filled with ether, any object moving in space is naturally affected by its presence. FitzGerald and then Lorenz each proposed that the *"..dimensions of solid bodies are slightly altered by their motion through the ether."* In the same manner that a body moving in air encounters the resistive pressure of air compressing the body, a body moving in ether should encounter a resistive pressure of ether. As a result, it should contract in the direction of motion.

Lorenz theorized that a body moving through ether does not drag ether. However, Michelson's experiment showed that the earth drags ether. Lorenz, in order to save his theory, suggested that the earth moving through ether might contract, only 6 centimeters,[158] in the direction of its motion and the contraction could explain the result of Michelson's experiment without the assumption that the earth drags ether.

Based on these ideas, Lorenz developed his relativity theory and the relativity of length formula, by which one could calculate how much a given velocity relative to ether would change the length of a body.

Joseph Larmor, in 1900, had reached the conclusion that a clock moving with velocity "v" relative to ether, must

---

[158] - Collected Papers of H. A. Lorenz, Volume IV, p. 221. Martinus Nijhoff, 1937.

run more slowly than it would when stationary.[159] Lorenz made a similar prediction. Since a moving clock experiences a pressure of ether inside the clock, the oscillating parts, including the electrons orbiting the nuclei, also experience the resistive pressure of ether and are less free to circle, oscillate or move. As a result, the clock will run more slowly.

In 1904, Lorenz published his formula for the relativity of mass.[160] Here follows a summary of his mathematical reasoning:[161] a moving body experiences the resistive pressure of ether. The higher the speed of the body, the greater is the resistance force of ether. Using the formula $F = m\alpha$, Lorenz concluded that a moving body must have greater mass than it would while stationary.

In June 1905, Poincaré published several proposals concerning the relativity theory.[162]

Three months later, in September of that same year, Albert Einstein published a paper on relativity[163] which proposed the same ideas as those of Lorenz and Poincaré, but disguised. Many scientists, realising that Einstein's paper duplicated those of Lorenz and Poincaré', criticised Einstein and his paper. In order to silence criticism Einstein published three books, namely "*The Meaning of Relativity*", "*The*

---

[159] - Larmor, *Aether and Matter*, (1900).

[160] - *Amst. Proc.* vi (1904), p. 809.

[161] - Since $F = ma$, (force = mass x acceleration) with a constant acceleration, then if the force is to increase, mass also must increase. A body experiencing the resistive pressure of ether needs a greater force to keep its acceleration constant. A greater force means a greater mass.

[162] - Henri Poincaré, "Sur La Dynamique de L'Electron", *L'Academie des Sciences*, t. 140, (5 Juin 1905), p. 1504-1508.

[163] - *Ann. d. Phys.* xvii (Sept. 1905), p. 891. See Appendix for a copy of English translation of the relativity paper called "Electrodynamics of Moving Bodies" by A. Einstein.

Evolution of Physics" and *"Relativity"* to explain his 1905 relativity paper and to show that it differed from those of Lorenz and Poincaré.

The following is how Einstein, subsequently in his books, explains his relativity theory. Just as a magician fools observers by creating an illusion of reality, similarly Einstein was able to fool many by presenting Lorenz's relativity theory in alternate form. Just as a magician, by using light in a special way, can deceive people into believing something is true, similarly Einstein, by using the speed of light, was able to deceive mankind into believing that his ideas were true.

To see the difference between Lorenz' theory and Einstein's ideas, let us compare the fundamentals of Lorenz' theory with those of Einstein. In Lorenz' theory, there is a real contraction of length; in Einstein's theory, the contraction is illusory. In Lorenz' paper, the cause of the contraction of length, the mass increase and the slowing of clocks, is the ether that exists in space; whereas in Einstein's paper, the cause is the speed of light.

Einstein explains his reasoning with an example:[164] two light signals from two simultaneous lightning strokes, one from A and the other from B, move towards an observer who is sitting stationary exactly in the middle of two points. The two signals reach the observer at exactly the same time; therefore the two events (the lightning) for the stationary observer are simultaneous. If the observer is not stationary, but is moving toward B, then the light signal from B reaches the observer sooner than from A. Therefore, for the moving observer, the two events are not simultaneous. Einstein then concludes that the events which are simultaneous with

---

[164] - See Einstein, *Relativity, the Special and the General Theory*, Methuen & Co. Ltd., London, see Sec. IX or pages 25-26.

reference to a stationery observer are not simultaneous with respect to a moving observer.[165]

Einstein replaces actual time with the time light from the event reaches the observer. In other words, he says that the time an event occurs is the moment the light reaches the observer; the actual time is not a reality. What counts is the time the observer sees. If a star is located one light year away and if it happened that one year ago it exploded and we are now seeing the explosion, according to Einstein, the star is exploding now. That is to say, the difference in location or distance means nothing. Furthermore, there is another problem in the way Einstein presents his ideas about motion and simultaneity. Einstein hides the fact that any stationary observer who happens to be standing close to B would also see the signal from B faster than from A. This means that not only for moving observers but also for all stationary observers who happen to be standing close to one of the events, the two events are not simultaneous.

Regarding the relativity of length, Einstein claims that a body moving in empty space contracts in the direction of motion because of the speed of light.[166] To summarise this, if we want to measure the length of a stationary object, we measure one end of the object to the other end against a measuring rod and determine its length. If the object is moving, the length will appear to be shorter. The reason is this: if we first look at one end of the moving object against a measuring rod and then the other end to see how long it is, by the time light moves from one end to the other (Einstein's

---

[165] - See Appendix for a copy of the section "The Relativity of Simultaneity" taken from Einstein's book, *Relativity*.

[166] - Albert Einstein, *Relativity, the Special & the General Theory*, Translated by R. W. Lawson, 3rd ed, Methuen & Co. Ltd. London, (1920), p.28.

definition of simultaneity), that end meanwhile has moved a distance. As a result, the object appears to be shorter. In brief, a moving rod will appear to be shorter than when it is stationary, a reasoning that has been proven to be misleading. P. Painlevé has shown that if the direction of motion is reversed, instead of contraction there is elongation.[167] Furthermore, Einstein mixes the apparent length with the actual length and deceives people into believing that the motion of a rod causes the actual length to contract. What does the apparent length have to do with the actual length? A long rod viewed from its end will appear as a point. What does the appearance of the rod have to do with the actual length of the rod? In Lorenz' theory, a body moving in ether contracts in the direction of motion by the resistive pressure of ether, and the contraction is real, whereas in Einstein's paper, the contraction is only apparent and not real. In Lorenz's theory the motion always causes the contraction and the contraction does not depend on which direction the body is moving, but in Einstein's paper, instead of contraction there could be elongation, depending on which direction the body is moving. Einstein's reasoning has been simplified by using the following example:

Suppose we have a bar placed alongside a measuring rod.

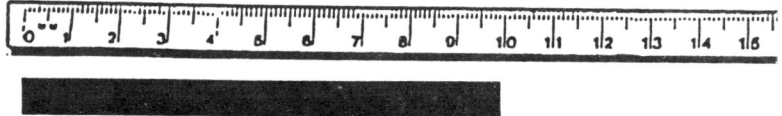

Fig. 1

---

[167] - P. Painleve, an eminent French mathematician, has disproved Einstein's theory by showing that if the direction of motion is reversed, instead of contraction we have elongation. See *Science Abstracts*. Sec. A-Physic: vol. 25, part 3, (March 31, 1922), p. 170.

If both the bar and the measuring rod are stationary (Fig. 1), the length of the bar can easily be determined by superimposing one end of the bar on the zero reading and the other end of the bar that coincides with a number, (let us say 10 meters) on the measuring rod, indicating the length of the bar in meters. However, if the measuring rod is stationary but the rod is moving to the left (Fig. 2), in this case we will not be making the same exact measurement as before and the bar appears to be shorter. At the instant that one end of the bar coincides with the zero indicator on the measuring rod, we make a note of it, and by the time (as fast as the speed of light) we turn our attention to the other end of the bar, in order to determine its length, the bar meanwhile has moved and its end has dislocated a certain distance to the left. As a result we make a reading that is less than 10 meters.

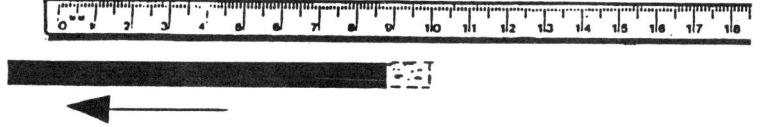

Fig. 2

This means that the bar, whose length in case one was measured to be 10 meters, now appears to be less than 10 meters. Based on these kinds of examples Einstein claimed that the actual length of bodies in motion is shorter than when they are stationary.

In the examples that Einstein gives, the direction of motion of the bar in relation to the frame of reference (measuring rod) was chosen in such a way as to show a contraction. However, there is a fundamental flaw in this kind of reasoning, because if the direction of motion is reversed, we will have elongation instead of contraction. For example, in the case before, instead of moving to the left, if the rod moves to the right (Fig. 3), then the length of the bar appears to be longer.

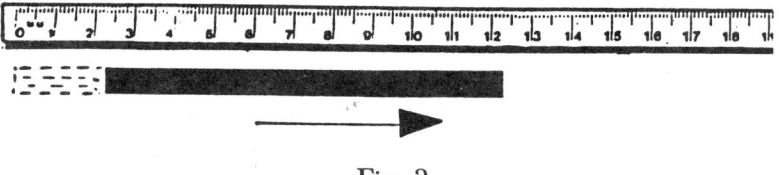

Fig. 3

This is because if at the instant that the one end of the bar coincides with the zero reading and we make a note of it, by the time we turn our attention to the other end of the bar, in order to determine its length, the bar has moved a distance to the right and its end coincides with a number which is more than 10 meters. Here we see that by changing the direction of motion of the bar our measurement, instead of showing a contraction, shows elongation. The fact that instead of contraction there could be elongation proves that Einstein's ideas have another side to them that he avoided mentioning otherwise it would have proven that the whole idea is wrong.

According to Lorenz' theory, a moving clock runs more slowly than when it is stationary because the movement of the clock is slowed down by the resistive pressure of ether. In the same way, a wrist watch without its case, if moved very fast in open air, definitely either slows down or stops as air pressure interferes with its movements. Similarly, the pressure of ether (magnetic pressure) interferes with the movements of a clock and slows it down.

However, according to Einstein's interpretation, a moving clock runs more slowly because time itself slows down. Einstein's reasoning[168] is this: he uses the formula x = vt (distance equals velocity multiplied by time). Having already confused the apparent length with the real length, meaning that an increase in velocity decreases the real distance (length, x), he then, using the above formula, arrives at the idea that time must slow down. What does the speed of light have to do with the time of a clock?

According to Lorenz' paper, a body moving in ether behaves as if its mass were increased, because the body encounters the resistive (magnetic) pressure of ether. The higher the speed the higher will be the resistive force of ether. For this reason, a moving body needs a greater force for its acceleration than a stationary one. Using the formula $F = m\alpha$, Lorenz concluded that a moving body behaves as if its mass were increased.

In Einstein's paper, however, he claims that a body in motion increases its mass from a space devoid of any matter, that is to say, a body gains mass out of absolutely nothing. Einstein's reasoning for the mass increase, again, is the speed of light. This reasoning has been shown to be faulty[169] on these grounds: Einstein assumes that a contraction in the apparent length causes an actual increase in density. He then mixes the density with the actual mass and assumes that an increase in density means an increase in actual mass.

It must be noted that the proof of the formulas originally belonging to Lorenz' theory, Einstein by misrepresentation claimed as proof of his new version of the

---

[168] - *Ann. d. Phys.* xvii (Sept. 1905), p. 891. - The English translation taken from: *Introduction to the Special Theory of Relativity*, Claude Kacser, Department of Physics and Astronomy, University of Maryland, Prentice-Hall, Inc., Englewood Cliffs, New Jersey, p. 49, 4.

[169] - Otto Luther, *Relativity is Dead*, California, Key Research Corp., (1986).

relativity theory. Many scientists realised that the paper misrepresents facts and theories but they were powerless to do anything about it because, as stated previously, Einstein's aid, Minkowski, made all things seem possible by bringing in the fourth dimension. Although many scientists have written books about the problems with Einstein's papers, all have been ignored. Consequently, for almost a century, Einstein has misled mankind, who has been so intrigued with ideas of the fourth dimension and time travel that it has accepted, unexamined, absurdities which contravene many laws of physics and do not even agree with common sense.

Today when one studies all the so-called 'proofs' of the Einstein version of relativity, one sees that they are based on formulas originally developed on the basis of ether by Lorenz; and that they are actually proofs of Lorenz' relativity, and not that of Einstein's. However, Einstein in his books, by misrepresentation and distortion of facts, has created a false idea about ether and relativity. He created the misconception that since relativity has been proven to be correct, therefore ether has been disproved.. In other words, he used relativity as evidence against ether; yet one of the fundamental characteristics of ether is relativity, that is, the length and mass of a moving body are affected by ether and a moving clock will run more slowly by the effect of ether. Einstein, by his misrepresentations in the name of modern physics, has led theoretical physics down a blind alley, effectively retarded scientific progress and slowed down the pace of advancements in science. Only in the future can mankind estimate the enormity of the damage he has caused.

Einstein's claims that all motion is relative and that absolute time and space are non-existent, are also evidence of misrepresentation

The following is an excerpt taken from Charles Lane Poor's book, <u>Gravitation Versus Relativity</u>, which, in a few pages, illustrates how deceptive and misleading is the basis of Einstein's ideas.

> *Einstein asserts that the Michelson-Morley experiment is final and conclusive, and he explains the result of the experiment by the assertion that there is no "absolute motion", there is no "absolute space", there is no "absolute time". All motion is relative: the steamer moves relatively to the earth, the earth moves relatively to the sun, the sun relatively to the stars. Nothing exists independently of the observer; all is relative, nothing is absolute. Hence the name:– RELATIVITY THEORY.*
>
> *This general postulate, or assumption, of the relativity of all motion and the non-existence of absolute motion is explained, illustrated and enforced by Einstein in the following way. He supposes two observers, one in a railroad train running on a straight stretch of track at uniform speed, the other observer standing beside the track and watching the train go by. Just as the train passes the watcher on the ground, the person in the train leans out of the window and drops a stone. This stone partakes of the forward motion of the train, and to the person who let it fall it appears to fall in a straight line, as shown in the accompanying diagram. But to the watcher on the ground, the stone, as it falls toward the earth, appears to move*

forward in the same direction as the train; to him it appears to describe a curved line, a parabola.

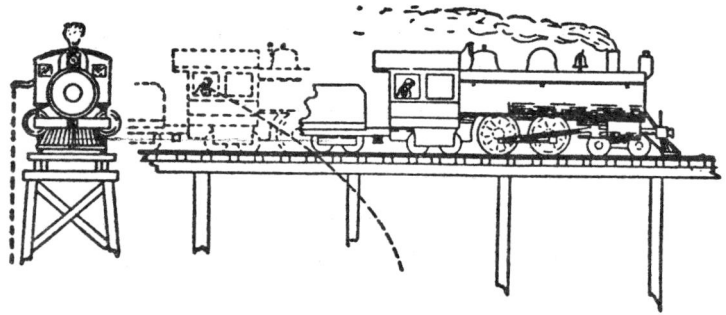

Fig. 1

Which, if either, is the true path of the stone: the straight line as it appears to the one observer, the parabola as the other sees it, or is it some other curve, compounded of these and the motion of the earth? To Einstein the answer is simplicity itself, the stone has no path. "With the aid of this example it is clearly seen that there is no such a thing as an independently existing trajectory (lit. "path-curve"), but only a trajectory relative to a particular body of reference"(10).[170]

What does this statement of Einstein mean? The stone certainly left the window of the car and came to rest at some point on the ground at the side of the track. How did it get from the window to the ground? The ordinary common-sense answer

---

[170] - Albert Einstein, *Relativity, the Special & the General Theory*, Translated by R. W. Lawson, 3rd ed, Methuen & Co. Ltd. London, (1920), p. 10.

would be that the stone travelled in some curved path through space, from one point to the other. It is true that the path might appear differently to different observers; as a straight line to a person in the train, as a parabola to a watcher on the ground, as a twisted curve to an aviator flying diagonally over the train: but no matter how the path appeared to these various observers, the stone travelled in one single definite path; it would have travelled the same path, if no one had watched it fall. Now this simple common-sense statement, that the stone actually did pass from the one point to the other in some definite path, is exactly what Einstein, in his statement above quoted, denies. He states that the stone had no path independent of an observer, that the path it travelled depended upon the person who watched it fall, that it actually had different paths for different observers . . .

This postulate of relativity is the denial of the existence of any reality behind our observations. The physical, the material world of land and water, of trees and houses, of men and women does not actually exist; it is not a real world. It exists only in and through the observer, and is different for different observers. The path of a falling stone is a straight line for one observer, a parabola to another; a steel rod is a yard long to one person, and a different length to another; each and every observer is correct; the stone has no "true" path, the rod has no "real" length. The fantastic picture of the cubist is as true to nature as the work of a Corot or a Meissonnier.[171]

To emphasise how right is Professor Poor's argument, and how misleading is Einstein's idea, let us use the example that Einstein has given. Let us assume that an observer is standing on the roof of the train, exactly above the hand of the person releasing the stone. The observer on the roof cannot see the motion or displacement of the stone as it is falling towards the ground. What Einstein is saying is that because he does not see the stone moving, there is no displacement at all, and he is just as correct as the observer on the ground. This is a clear misrepresentation and denial of reality. The fact is this, the person on the roof sees the stone becoming smaller, and immediately understands the reality that the stone is getting farther away, and therefore must be falling. Einstein is saying that the optical illusion is reality, which is absurd. This implies that a magician's feats are also reality, because reality is whatever an observer sees. Consider a cross-eyed observer, who sees two of everything. According to Einstein, there must then be two of everything. If the observers close their eyes, everything ceases to exist. This is the essence of Einstein's idea. No wonder Rutherford called it a 'joke' and Soddy and Essen, among others, called the whole business a "swindle".[172]

---

[171] - Charles Lane Poor, *Gravitation Versus Relativity*, (1922), G. P. Putnam's Sons, New York, London, p. 19-21.

[172] - Louis Essen, *Relativity - Joke or Swindle,* Electronic and Wireless World, v. 94, (Feb. 1988), p. 126-127.

Einstein and the Fraud of the Century

In what follows, you will find a critical examination of the first few pages of Einstein's relativity paper, with some ensuing conclusions. Let us examine Einstein's paper without regard to its fame or any preconceived ideas which, through the power of suggestion, have been created in the minds of the generality of mankind. In this light, free from prejudice, our own eyes may discern the truth from the facts as they present themselves.

To understand Einstein's method of reasoning and how his assumptions and conclusions were made, the first few pages of the theory have been presented here. These pages, which are suppose to make Einstein's paper different from that of Lorenz and Poincaré's, give some clues about the rest of his paper. Upon reading Einstein's work for the first time, some may find its contents difficult to understand; however, if you read carefully the simple explanations and illustrations provided in this book, you will notice that you will be able to understand the essence of Einstein's work without any difficulty.

"ON THE ELECTRODYNAMICS OF MOVING BODIES
By A. EINSTEIN[173] "

It is known that Maxwell's electrodynamics - as usually understood at the present time - when applied to moving bodies, leads to asymmetries which do not appear to be inherent in the phenomena. Take for example, the reciprocal electrodynamic action of a magnet and a conductor. The observable phenomenon here depends only on the relative motion of the conductor and the magnet. Whereas the customary view draws a sharp distinction between the two cases in which either the one or the other of these bodies is in motion. For if the magnet is in motion and the conductor at rest, there arises in the neighbourhood of the magnet an electric field with a certain energy, producing a current at the places where parts of the conductor are situated. But if the magnet is stationary and the conductor in motion, no electric field arises in the neighbourhood of the magnet. In the conductor, however, we find an electromotive force, to which in itself there is no corresponding energy, but which gives rise - assuming equality of relative motion in the two cases discussed - to electric current of the same path and intensity as those produced by the electric forces in the former case.

Examples of this sort, together with unsuccessful attempts to discover any motion of the earth rela-

---

[173] - Claude Kacser, *Introduction to the Special Theory of Relativity*, Department of Physics and Astronomy, University of Maryland, Prentice-Hall, Inc., Englewood Cliffs, New Jersey, p. 49, 4.

tively to the "light medium" suggest that the phenomena of electrodynamics as well as of mechanics possess no properties corresponding to the idea of absolute rest. They suggest rather that, as has already been shown to the first order of small quantities, the same laws of electrodynamics and optics will be valid for all frames of reference for which the equations of mechanics hold good.* We will raise this conjecture (the purport of which will hereafter be called the "principle of relativity") to the status of a postulate, and also introduce another postulate, which is only apparently irreconcilable with the former, namely, that light is always propagated in empty space with a definite velocity c which is independent of the state of the motion of the emitting body. These two postulates suffice for the attainment of a simple and consistent theory for the electrodynamics of moving bodies based on Maxwell's theory for stationary bodies. The introduction of a "luminiferous ether" will prove to be superfluous inasmuch as the view here to be developed will not require an "absolutely stationary space" provided with special properties, nor assign a velocity-vector to a point of the empty space in which electromagnetic processes take place.

The theory to be developed is based - like all electrodynamics - on the kinematics of the rigid body, since the assertions of any such theory have to do with the relationship between rigid bodies (systems of co-ordinates), clocks, and electromagnetic processes. Insufficient consideration of this circumstance lies at the root of the difficulties which the electrodynamics of moving bodies at present encounter."

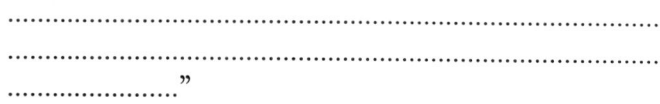

* - *The preceding memoire by Lorenz was not at the time known to the author.*

Einstein in the second, third and fourth paragraphs gives two examples: one, the example of a conductor and magnet, and the other, the unsuccessful attempts to discover the effects of the motion of earth relative to ether, which is the result of Michelson's experiment (see page 149). Based on these two examples, he invents a postulate, a principle and some conclusions totally unrelated to the examples which are only a set-up. The following is a summary of the two examples followed by his postulates and suggestions.

Examples:

1. ... *reciprocal electrodynamic action of a magnet and a conductor. The observable phenomenon here depends only on the relative motion of the conductor and the magnet...* "

2. . . . *the unsuccessful attempts to discover any motion of the earth relatively to the "light medium,* [ether] *suggest:*

that:

A. *the phenomena of electrodynamics as well as of mechanics possess no properties corresponding to the idea of absolute rest.*

and also suggest the postulates:

214

B. ...*The same laws of electrodynamics and optics will be valid for all frames of reference for which the equations of mechanics hold good.*

C. *. . . . light is always propagated in empty space with a definite velocity c which is independent of the state of motion of the emitting body.*

Here we are going to study each of the above examples to see if these suggestions are true or not. Einstein derives his relativity assumptions using the example of a magnet and a conductor, and from the reciprocal action or relative motion between the two. To see what is involved here, let us take the reciprocal action of two gears positioned in such a way that when one of the gears is turned, the other one also turns, but one turns in one direction, and the other turns in the opposite direction (Fig. 1). This shows that there is a reciprocal action between the two, that is, the motion of one affects the other.

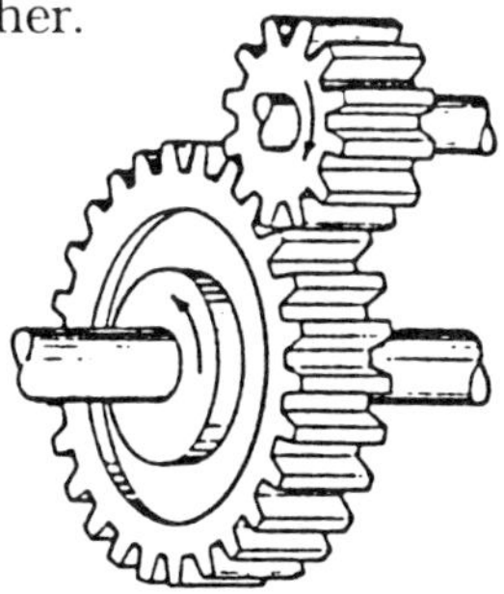

Fig. 1

Let us use another example. Let us take the reciprocal action between two fans which is very similar to the reciprocal action between a magnet and a conductor. If we place two electric fans very close but opposite to each other, and if one is turned on, the force of its wind will affect and turn the propeller of the other fan, as if both fans were turned on. If the fan which was connected to electricity is turned off and if its propeller is made to stop turning at once, the force of the wind of the other fan will cause it to turn again. How could examples of this kind be considered as evidence to suggest that:

> *The same laws of electrodynamics and optics will be valid for all frames of reference for which the equations of mechanics hold good.*

(What does the reciprocal action have to do with the laws of optics?)
    Or   *...the phenomena of electrodynamics as well as mechanics possess no properties corresponding to the idea of absolute rest.*
    Or *. . light is always propagated [travels] in empty space with a definite velocity c. . . "*

What does reciprocal action have to do with any of the above suggestions? There is no connection between his example and his conclusion.
    Moreover, consider Einstein's statement:

> *Take for example the reciprocal action of a magnet and a conductor. The observable phenomenon here depends* only on *the relative motion of the conductor and the magnet. . .*

Here it will be shown that the observable phenomenon does not depend only on the relative motion of the two, but also on other variables. Let us take a first case where the conductor is at rest and the magnet is in motion. Let us assume a current equal to .01 ampere is generated in the conductor. In a second case, the magnet is at rest and the conductor in motion. With a little attention, one realizes that the current in the second case could be different from that of the first case, for one simple reason: the motion of the conductor in the magnetic field of the earth could cause an additional current in the conductor. This shows that the amount of current in the conductor does not necessarily depend solely on the relative motion of the magnet and the conductor, which clearly shows that Einstein's statement: "...the.... *phenomenon here depends only on the relative motion of the conductor and the magnet,*" is not true.

The second example was based on "...*the unsuccessful attempts to discover the effect of the motion of earth relatively to the light medium,*" which was the result of the Michelson and Morley experiment. This example, according to Einstein, suggests the following:

> A. "*the phenomena of electrodynamics as well as mechanics possess no properties corresponding to the idea of absolute rest.*"
> B. "*The same laws of electrodynamics and optics will be valid for all frames of reference for which the equations of mechanics hold good.*"
> C. "... *light is always propagated in empty space with a definite velocity c...*"
> D. [The velocity of light] "...*is independent of the state of motion of the emitting body.*"

In previous chapters, it was shown that the idea of looking for the wind created by the motion of the earth through ether, on or near the surface of the earth, is totally wrong. It was shown that the earth drags not only its air atmosphere but also drags ether. Furthermore, it was proven that stellar aberration is clear evidence that the earth drags ether. It was shown that the motion of the air atmosphere of the earth itself must accompany ether. To think that the earth drags its air atmosphere and not the particles that fill in the spaces between the molecules of air is totally absurd. Finally, it was shown that Michelson's experiments actually prove the earth drags ether.

Everyone knows that inside a moving car with the windows shut, air is dragged by the car. As a result, there is no wind blowing on the passengers and no motion of the car relative to the air inside the car. No matter where the source of sound is, whether it is located inside or outside the car or whether the car is moving or stationary, approaching or moving away, the speed of sound reaching the passenger's ear is always the same. This is exactly what Michelson's experiment showed with light instead of sound, namely, that the speed of light, regardless of the earth's speed or direction, is always the same. Michelson's instrument registered no 30 kilometres per second ether wind, which, in turn, proved there is no motion of the earth relative to the ether (light medium) surrounding the earth.

Today, the discovery of the solar wind behind the magnetic bow shock at 10,000 kilometres above the surface of the earth proves that the wind that Michelson was looking for exists beyond the limit of the bow shock, and not at the low heights of a few kilometers. This further explains why Michelson did not find the ether wind.

How could this simple and logical explanation be ignored and instead interpreted such that:

A. "*The phenomena of electrodynamics as well as mechanics possess no properties corresponding to the idea of absolute rest.*"

B. "*...the same laws of electrodynamics and optics will be valid for all frames of reference for which the equations of mechanics hold good.*"

C. "*. . . light is always propagated [travels] in empty space with a definite velocity. . .*"

D. [The velocity of light] "...*is independent of the state of motion of the emitting body.*"
E. "*Light travels in empty space*" [without the medium of ether]. See the postulate.

How could such a phenomenon that has such a simple and obvious explanation become the basis of such unreasonable assumptions that defy many laws of physics?

To see how unreasonable Einstein's suggestions are, let us study the postulate that Einstein has fabricated which concerns light and ether. First, Einstein proposes that space is empty by his postulate that light travels with a definite velocity in empty space. On the one hand, if we assume light is not a wave of ether, but rather something that travels in empty space, then there is no reason why light should always travel with a definite velocity. There is no reason why light does not slow down. Such an assumption defies the known laws of nature. On the other hand, if we consider that space is not empty but is filled with ether and that light is a wave of ether, then we see that one of the fundamental characteristics of any wave is that it travels with a definite velocity, a velocity that depends on the nature of the medium. Furthermore, another fundamental nature of a wave is that its velocity does not depend on the motion of the source. For example, the speed of sound does not depend on how fast the source is moving. If it is approaching or moving away, the speed of sound is always independent of the motion of the source because the sound wave is carried by the medium of air and not by its source. Einstein, giving no reason or evidence, states that light travels in empty space and its velocity does not depend on the state of the motion of the emitting body. Einstein, without giving a single reason, has fabricated a postulate that contradicts the basic laws of nature. It must be noted that Einstein fabricated the postulate only to disprove ether and set up the stage for his relativity theory.

Furthermore in the second paragraph Einstein claims:

> *The introduction of a "luminiferous ether" will prove to be superfluous inasmuch as the view here to be developed will not require an "absolutely stationary space"....*

When one carefully studies the claim, one can see that not only is it false, but incomprehensible. How can the proof of the existence of ether be related to absolutely stationary space? What does the proof of any medium like air have to do with stationary or moving space? The thought is so confusing that no one can make sense out of it. Furthermore, the postulate that he himself fabricated assumes the existence of absolutely stationary space. The postulate says: *"...light is always propagated in empty space with a definite velocity c".* If empty space is not stationary how can light have a definite velocity? Then on what basis does Einstein claim that his theory will not require an absolutely stationary space? Can space be other than stationary?

Einstein claims that his *"...theory of electrodynamics of moving bodies is based on Maxwell's theory for stationary bodies."* The idea is further evidence of misrepresentation, because Maxwell did not have a relativity theory for Einstein to base his theory upon. In the same paragraph, Einstein fabricates the idea that electromagnetic processes take place in empty space by claiming that

> *...the view here to be developed will not require an 'absolutely stationary space' provided with special properties, nor assign a velocity-vector to a point of*

the empty space in which electromagnetic processes take place. [174]

Later on in this book, it will be demonstrated that Einstein misrepresented Maxwell's electromagnetic theory to disprove the existence of ether.

There are many examples of these sorts of misrepresentations throughout his papers, too many to be mentioned.

These facts give you some idea about the others. Furthermore, Einstein, for no apparent reason, applies Lorenz' relativity formulas to different cases, such as Maxwell's equations, and reaches totally unrelated conclusions. These are exactly the reasons why Arthur Lynch wrote:

> . . . *Then Einstein always a copier and borrower adapts the formula of Lorenz to cases where it has no application.*[175]

This explains how Einstein, a patent officer, was able to write in one year his relativity paper and twenty others, as well as a Doctoral thesis, all of which apparently no one could understand.

---

[174] - For further detail see Appendix G, "Evidence of Fraud: Einstein's Misrepresentation of Maxwell's Theory."

[175] - Arthur Lynch, *The Case Against Einstein,* p. 70.

Einstein's Misrepresentation of Maxwell's Theory: Evidence of Fraud

Before Einstein, the general understanding of all scientists was that Maxwell's electromagnetic theory was based on ether and that, according to the theory, light is the wave of the medium of ether. For example, the 1890 edition of the Encyclopaedia Britannica under the section "Ether" (written by Maxwell himself, summarising his theory), explains that light is propagated by a medium called ether.[176] The general understanding of scientists concerning Maxwell's theory (namely, that it is based on ether), lasted some time after Einstein. This can be seen in many scientific books including some of the encyclopaedias. For example, the 1920 edition of the American International Encyclopedia, writes:

> *The electromagnetic theory of light propounded by James Clerk-Maxwell in 1865, assumes that electricity is an elastic displacement or tension in ether and that magnetism is the inertia of the ether.*[177]

The 1967 edition of the Encyclopaedia Britannica, writes:

> *Maxwell was able to show that all electromagnetic and optical phenomena could be explained by a single system of stresses in the ether.*[178]

---

[176] - See *Encyclopaedia Britannica*, (1890), vol. VIII, p. 570.

[177] - *American International Encyclopedia*, New York, (1950), Vol. VI, see "Ether".

[178] - *Encyclopedia Britannica*, 1967, vol. 8, p. 749.

However, Einstein described Maxwell's theory quite differently. According to Einstein, Maxwell's electromagnetic theory has nothing to do with ether, rather, that light is simply electromagnetic waves and that: "*The electromagnetic wave spreads in empty space.*" [179]

Here one can see two different versions of Maxwell's electromagnetic theory . In order to see which version is correct, I myself studied Maxwell's actual theory,[180] and found that Maxwell had clearly based his theory on the existence of the medium of ether. He had conjectured that electric and magnetic fields are those areas where ether is in motion; that electric and magnetic forces are created from the motion of ether; and that light is the vibration of ether. Comparing Maxwell's paper with Einstein's papers and books,[181] it becomes clear that Einstein has misrepresented Maxwell's theory.[182] For detailed information, see Appendix I.

Further consideration of all of the foregoing led me to raise my strong suspicion that Einstein's misrepresentation of Maxwell's theory is only one of a series of clear, irrefutable indications of fraud. Einstein, by distorting facts and manipulating theories, was able to mislead mankind into believing there is no ether, while at the same time 'prove' his illusive relativity theory.

---

[179] - See Einstein and Infield, *The Evolution of Physics,* Simon and Schuster, New York, (1938), p. 154-155.

[180] - See *Philosophical Magazine and Journal of Science*, Dec. 1864, p. 460.

[181] - See Einstein, *Relativity, the special and General Theory,* fn. 24. See also Einstein & Infield, *The Evolution of Physics*, Simon and Schuster, New York, (1938), p. 129-157.

[182] - For more information, see Appendix, "Evidence of Fraud: Einstein's Misrepresentation of Maxwell's Theory".

Up to the time of Einstein, no scientist doubted the existence of ether. Only after Einstein had misrepresented facts and fabricated a series of false ideas about ether did scientists begin to doubt its existence.

Light being the vibration of ether is a very simple concept that can explain all the related phenomena in nature without difficulties or contradictions. However, Einstein discarded this simple and logical concept and replaced it with two contradictory and illogical theories, namely the photon theory and his perversion of Maxwell's electromagnetic theory. Einstein himself admitted that these two theories are contradictory and difficult to reconcile.[183]

---

[183] - Albert Einstein and Leopold Infeld, *The Evolution of Physics*, Simon and Schuster, New York, (1938), p.295.

EINSTEIN'S MISREPRESENTATION OF HERTZ'S RADIO WAVE: EVIDENCE OF FRAUD

Let us see what Hertz, who discovered radio waves and invented the radio, understood from Maxwell's electromagnetic theory. Was Hertz' discovery of the radio wave based on ether or an electromagnetic wave in empty space? After careful study of Hertz' papers, I found that Hertz's conviction of the existence of ether is unquestionable. The following are Hertz' own words concerning light and ether:

> *What then is light? Since the time of Young and Fresnel we know that it is a wave-motion. We know the velocity of the waves, we know their wavelength, we know that they are transversal waves; in short, we know completely the geometrical relations of the motion. To the physicist it is inconceivable that this view should be refuted; we can no longer entertain any doubt about the matter. It is morally certain that the wave theory of light is true, and the conclusions that necessarily follow from it are equally certain. It is therefore certain that all space known to us is not empty, but is filled with a substance, the ether, which can be thrown into vibration . .[184]*

In another talk he says;

---

[184] - Miscellaneous papers by Henrich Hertz with an introduction by Professor Phillop Lenard, (1896), p. 313, 314.

> . . . remove from the world the luminiferous ether, and electric and magnetic actions can no longer traverse space.[185]

In contrast to the above, Einstein describes[186] radio waves to be electromagnetic waves that travel in empty space. In Einstein's books, the views and understanding of Hertz about radio waves are misrepresented in such a way as to disprove the existence of ether while proving Einstein's ideas.

---

[185] - Albert Einstein & Leopold Infield, *The Evolution of Physics,* Simon and Schuster, New York, (1938), p. 152.

[186] - See Einstein & Infield, *The Evolution of Physics,* fn. 183, p.129 and 156.

Einstein's use of Faraday's Old Ideas: Evidence of Misrepresentation

Over the course of his life, Faraday came to believe in the existence of ether. However, in the early stages of his scientific investigations, his discussion of ether was mere conjecture. He wondered about the function of ether in electromagnetic phenomena, and, in particular, wondered whether the ether which is the agent in the propagation of light might also be the agent in creating the magnetic or electric forces. He writes:

> *For my own part, considering the relation of a vacuum to the magnetic force, and the general character of magnetic phenomena external to the magnet, I am much more inclined to the notion that in the transmission of the force there is such an action, external to the magnet, than that the effects are merely attraction and repulsion at a distance. Such an action may be a function of the æther; for it is not unlikely that, if there be an æther, it should have other uses than simply the conveyance of radiation.*[187]

To validate this conjecture that the plays a role in both light and magnetism, Faraday after years of perseverance, was able to prove that his magnets could turn the plane of polarisation of a light beam through a certain angle. He also proved that the direction of rotation of the beam of light depends upon the polarity of his magnet. Faraday's discovery further proved the existence of ether.

---

[187] - Experimental Researches, 3075.

In order to disprove ether and prove his own relativity ideas, Einstein took mankind back to Faraday's early, undefined conjecture and used it to support his own conjecture regarding magnetism, in which he avoids having to acknowledge ether by focusing on the term "magnetic field". He writes:

> *If, for instance, a magnet attracts a piece of iron, we cannot be content to regard this as meaning that the magnet acts directly on the iron through the intermediate empty space, but we are constrained to imagine - after the manner of Faraday - that the magnet always calls into being something physically real in the space around it, that something being what we call a "magnetic field.' In its turn this magnetic field operates on the piece of iron, so that the latter strives to move towards the magnet. We shall not discuss here the justification for this incidental conception, which is indeed a somewhat arbitrary one.*[188]

Einstein has acknowledged the impossibility of magnetic attraction taking place in empty space. He explains magnetism by claiming that the magnet itself "*calls into being something physically real in the space around it.*" Notice that in the earlier quotation Faraday rejects this idea of *attraction and repulsion at a distance*.

Compare Einstein's explanation with the ether-based explanation found in Chapter 7.

---

[188] - Albert Einstein, *Relativity, the Special & the General Theory*, Translated by R. W. Lawson, 3rd ed, Methuen & Co. Ltd. London, (1920), p. 63.

FURTHER EVIDENCE OF FRAUD

The following are further examples of misrepresentations found in Einstein's work which, when compiled together, give you some idea of how he was able to highjack science into rejecting ether, and become famous. Since we are mainly concerned with the truth about ether, evidence presented here is all related to ether.

- Charles Lane Poor and others have shown[189] that Einstein, in order to prove his ideas, has misrepresented Fizeau's experiment.[190] Is this not an evidence of fraud?

- George Stoke, studying the phenomenon of stellar aberration, came to the conclusion that the earth must drag ether. Why did Einstein use stellar aberration as a proof of his relativity theory[191] and against ether? Is this not an evidence of misrepresentation?

- Why have Michelson's experiments, indicating that the earth drags ether, been distorted by Einstein in such a way as if they disproved ether?[192] Is it not a fact that Michelson himself considered his experiment as proof of the

---

[189] - Charles Lane Poor, *Gravitation Versus Relativity*, (1922), G. P. Putnam's Sons, New York, London, p. 266 - 267.

[190] - Albert Einstein, *Relativity, the Special & the General Theory*, Translated by R. W. Lawson, 3rd ed, Methuen & Co. Ltd. London, (1920), p.38.

[191] - Ibid, p.49.

[192] - See Albert Einstein and Leopold Infeld, *The Evolution of Physics*, Simon and Schuster, New York, (1938), p. 183.

earth's drag of ether?[193] Is this not a clear indication of misrepresentation?

Einstein claims that the negative result of Michelson's experiment can be explained by his relativity theory and not by ether, which he claims needs a "specially favoured co-ordinate system:"

> ...*without such a thing as a "specially favoured" (unique) co-ordinate system to occasion the introduction of the ether-idea, and hence there can be no ether-drift, nor any experiment with which to demonstrate it.*[194]

This is one of innumerable misrepresentations and distortions of fact. Einstein never mentions the fact that the earth's drag of ether explains Michelson's experiment perfectly. There is no need for *a "specially favoured" (unique) co-ordinate system..."* which is nothing but Einstein's own scheming to disprove ether and to prove his own ideas.

Many scientists objected to Einstein's relativity theory on the basis that it is sheer nonsense to claim that a clock's ticking slow or fast, or the contraction of length of a body, or a mass increase, could be caused by the speed of light, for there is no cause and effect relationship between the speed of light and these phenomena. Einstein replies:

---

[193] - *Amer. Jour. Sci.* xxii (1881), p. 386; and *Phil. Mag.*, Ser. 5, Vol. 24, No. 151, Dec. 1887, p. 459.

[194] - Albert Einstein, *Relativity, the Special & the General Theory*, Translated by R. W. Lawson, 3rd ed, Methuen & Co. Ltd. London, (1920), p.53.

> *The theory of relativity is often criticized for giving, without justification, a central theoretical role to the propagation of light, in that it founds the concept of time upon the law of propagation of light. The situation, however, is somewhat as follows. In order to give physical significance to the concept of time, processes of some kind are required which enable relations to be established between different places. It is immaterial what kind of processes one chooses for such a definition of time. It is advantageous, however, for the theory, to choose only those processes concerning which we know something certain. This holds for the propagation of light in vacuo in a higher degree than for any other process which could be considered, thanks to the investigations of Maxwell and H. A. Lorenz.[195]*

Is not Einstein's reply a clear evidence of manipulation and distortion of fact? The knowledge or accurate measurement of the speed of light does not justify Einstein to invent a theory about mass increase, or how a clock keeps time, or the contraction of length of a moving body, which are not even related to the speed of light. Does this kind of reasoning justify Einstein's rejection of ether?

Einstein claims that the most important proof of his theory is Michelson's experimental results:

> *But all experiments have shown that electro-magnetic and optical phenomena, relatively to the earth as the body of reference, are not influenced by the*

---

[195] - Einstein, *The Meaning of Relativity*, Princeton University Press, (1922), Princeton, p. 31.

translational velocity of the earth. The most important of these experiments are those of Michelson and Morley, which I shall assume are known. The validity of the principle of special relativity can therefore hardly be doubted.[196]

Many scientists have objected to the way the results of Michelson's experiments have been presented in text books and scientific publications. As an example, Petr Beckmann writes:

> It is thus ironic that this experiment, which is perfectly consistent with four out of five theories discussed here, and which refuted nothing but the unentrained (or partially entrained) version of the ether theory, should be held up in textbooks as proof of Einstein's theory or disproof of classical theory.[197]

- Why have Michelson's experiments been distorted in such a way as to support only Einstein's version of relativity? Dr. Louis Essen, a well known scientist of the time, writes:

> Insofar as the theory is thought to explain the result of the Michelson-Morley experiment I am inclined to agree with Soddy that it is a swindle; and I do not think Rutherford would have regarded it as

---

[196] - Ibid, p. 29.

[197] - Petr Beckmann, *Einstein Plus Two*, The Golden Press, Boulder, Colorado, (1987), p. 39.

a joke had he realised how it would retard the rational development of science.[198]

According to the Practical Standard Dictionary, the word swindle means "to cheat and defraud grossly or deliberately." We must bear in mind that in order for Einstein to prove his relativity ideas, he had to disprove ether. For this reason, he made a series of misrepresentations to disprove ether and to show that relativity, that is to say, changes of length, mass and time, have nothing to do with ether, but rather, are caused by the the speed of light.

- Ives and Stilwell, concluded that an experiment they conducted supports Lorenz' and Larmor's theory of time dilation. Einstein misrepresented the experiment as a proof of his theory of relativity.[199] For this reason, the editor of a collection of Ives' papers wrote:

. . . Einstein allowed both scientists and laymen to come to believe that this experiment supported relativity and nothing else. Was this stealing? No, not exactly. It is better called passive kidnapping.[200]

---

[198] - Louis Essen, "Relativity - Joke or Swindle", *Electronics and Wireless World*, v. 94 (Feb. 1988), p. 126-127.

[199] - *The New York Times*, 27 April 1938, p. 25.

[200] - *The Einstein Myth and the Ives papers: A Counter-Revolution in Physics*, Connecticut, (1979), The Devin-Adair Company, p. 85.

Then why, in all scientific publications, has this experiment been misrepresented as a proof of only Einstein's version of the theory of relativity?[201]

Why did Einstein misrepresent the facts about the mechanical view of ether,[202] claiming the impossibility of ether as jelly-like? Is it not a fact that nineteenth century scientists who had viewed ether as jelly-like were very much convinced of its existence? [203]

Einstein's false information about scientists' mechanical view of ether is further evidence of fraud. In order to disprove ether, he writes that experiments have shown that light must be a jelly-like substance, and then fabricates false information saying that physicists had to make *"...artificial and unnatural assumptions..."* which finally shattered their belief in ether:

> *In order to construct the ether as a jelly-like mechanical substance, physicists had to make some highly artificial and unnatural assumptions. We shall not quote them here, they belong to the almost forgotten past. But the result was significant and important. The artificial character of these assumptions, the necessity for introducing so many of them quite independent of each other, was*

---

[201] - For details, refer to N. Rudakov, *Fiction Stranger than Truth*, Australia, Published by N. Rudakov, (1981), p. 140-141.

[202] - For further detail, see Chapter 4, Qualities of Ether.

[203] - See *The Encyclopedia Britannica*, (1890), Vol. VIII, p.570.

enough to shatter the belief in the mechanical point of view.[204]

Any encyclopaedia or scientific publication, will invalidate these words of Einstein. There were no such assumptions made by anybody. No one else but Einstein shattered scientists' belief in ether. It was Einstein who "killed ether". In the above quotation, Einstein pretends that he is not the one who is disproving ether, but points instead to physicists who, he says shattered belief in ether by their "*.. artificial and unnatural assumptions*". Einstein by this kind of misrepresentation was able to mislead mankind into abandoning ether.

Einstein, in order to prove his relativity theory, distorts the facts and claims that experiments have proven that his theory is correct. For example:

Bodies with velocities approaching that of light would offer a very strong resistance to external forces. In classical mechanics the resistance of a given body was something unchangeable, characterized by its mass alone. In relativity theory it depends on both rest mass and velocity. The resistance becomes infinitely great as the velocity approaches that of light.

The result quoted enables us to put the theory to the test of experiment

Then he claims that:

---

[204] - See Albert Einstein and Leopold Infeld, *The Evolution of Physics*, Simon and Schuster, New York, (1938), p.123.

> ...*the experiment shows that the resistance offered by these particles depends on the velocity, in the way foreseen by the theory of relativity. In many other cases, where the dependance of the resistance upon the velocity could be detected, there was complete agreement between theory and experiment.* [205]

This is a false claim. It is a fact that Lorenz originally advanced the theory on the basis of classical mechanics; that a body moving in ether encounters the resistive pressure of ether and according to his formulas, at velocities close to the velocity of light the resistance is infinitely great. Einstein claims: *"In classical mechanics the resistance.... was ... unchangeable, characterized by its mass alone."*

Is this not an evidence of fraud? Furthermore, Einstein never mentions which experiment when and where, nor does he mention the possibility that external forces such as electricity and magnetism could be created by the current of ether. A force created by a current can do work up to a certain velocity, namely that of the current.

Hertz' understanding of Maxwell's work,[206] which was based on ether[207] has been misrepresented by Einstein as if to exclude ether, making it support his own ideas. Is this not clear evidence of fraud?

---

[205] - Ibid, p. 205-207.

[206] - *Miscellaneous Papers by Henrich Hertz* with an Introduction by Prof. Philipp Lenard, (1896), p. 319.

[207] - For further details, see the section called "Einstein's Misrepresentation of Hertz's Radio Wave: Evidence of Fraud".

The photoelectric effect phenomenon actually supports the wave nature of light.[208] Einstein misinterprets it so as to support his own idea that light is a particle that moves without ether, and to prove that mass increase and contraction of length have nothing to do with ether, but rather with the speed of light. Is this not clear evidence of fraud?

Einstein claims that: "... *the yearly movement of the apparent position of the fixed stars resulting from the motion of the earth round the sun (aberration)*" supports his special theory of relativity,[209] yet he gives no explanation as to how. Is this not evidence of fraud? Anyone studying stellar aberration can readily see that relativity has nothing to do with aberration. What does aberration have to do with mass increase, or the contraction of length due to velocity, or how a moving clock keeps time?

Einstein claims that the Doppler effect supports his theory of relativity.[210] Is this not an evidence of fraud? We are familiar with the Doppler effect in sound waves. What has the Doppler effect to do with mass increase, the contraction of length or how a clock keeps time?

Why has the relativity theory, which was developed on the basis of ether, been misrepresented as if to disprove its existence? Is it not a fact that Lorenz, long before Einstein, arrived at the relativity formulas on the basis of ether?

---

[208] - See the photoelectric effect formula..

[209] - Albert Einstein, *Relativity, the Special & the General Theory*, Translated by R. W. Lawson, 3rd ed, Methuen & Co. Ltd. London, (1920), p. 48.

[210] - Ibid, p. 50.

Why has the atomic clock experiment, (see p. 237) which actually proves Lorenz' relativity theory, been misrepresented and claimed to be a proof of Einstein's ideas? Is it not a fact that Lorenz, a year before Einstein, had predicted that a moving clock would run slow?

Why has Einstein been credited for theories that originally did not belong to him?

Why has the formula, $E = mc^2$, which belongs to Poincaré, been attributed to Einstein? Is it not a fact that Poincaré developed this formula five years before Einstein? Why, despite the fact that Einstein's paper has been proven to be defective, and that from a scientific point of view is worth absolutely nothing, has Einstein been credited with the formula? Is this not a clear evidence of misrepresentation?

In 1901, four years before Einstein, W. Kaufmann found that an electron's mass increases with velocity.[211] This was later confirmed by others. A year before Einstein, Lorenz, in 1904, found the formula for the mass increase.[212] Then why, in all the scientific publications, has the mass increase been attributed to Einstein as if he were the one who predicted the phenomenon and found the formula for the mass increase? Is this not a clear evidence of misrepresentation?

---

[211] - Kaufmann, *Akademie der Wissenschaften* (Gottingen) Nachrichten, Mathematisch-physikalische, Klasse, 1901, p.143-155, and in 1902, p. 291-296.

[212] - H. A Lorenz, *Proc. Acad. Sci.* Amsterdam, 6 (1904).

The photon theory contradicts experiments[213] and has been proven to be wrong, yet Einstein brought it into the science of physics as a fact. Is this not evidence of misrepresentation? We must remember that light as a photon means there is no ether.

Einstein claims that: "*The verdict is always against the assumption of the ether carried by motion,*" because "*...the velocity of light does not depend on the motion of the emitting source,...*" and that "*...we must, therefore, give up the analogy between sound and light waves...*"[214]

Is this not clear evidence of misrepresentation? Is it not a fact that the velocity of sound does not depend on the motion of its source? Bear in mind that, for example, while a moving car blows its horn, the air around the horn is being carried by the car and the speed of its sound reaching someone's ear who is standing by the side of the road, is always independent of the motion of the car.

A study of the books and encyclopaedias that were published before Einstein indicates that the general understanding of all scientists was that Maxwell developed his theory on the basis of ether.[215] A comparison of this understanding with the one presented in the name of Maxwell in scientific publications, including the text books of today, reveals major fundamental differences between

---

[213] - For further details, see Appendix F on the photon theory.

[214] - Albert Einstein and Leopold Infeld, *The Evolution of Physics*, Simon and Schuster, New York, (1938), p.179.

[215] - J. Clerk Maxwell, "A Dynamical Theory of the Electromagnetic Field", *Philosophical Magazine*, Dec. 1864, p. 460, (4).

them.[216] Maxwell's original theory has been manipulated and misrepresentation in such a way as to support Einstein's ideas. Many scientists had already objected to this kind of misrepresentation. Was this discrepancy an accidental mistake or further evidence of fraud?

In order to do away with ether, Einstein claimed that the reflection of light can be explained by *corpuscles* or photon theory.[217] He compared corpuscles with elastic balls thrown against a wall and reflected back. This analogy does not stand up to analysis, because it is based on the fact that the balls are larger than the molecules of the wall. Corpuscles, however, would be much smaller than atoms; otherwise, we would be able to detect them. Such tiny particles would pass all the way through the wall without being reflected back, as the possibility of one colliding with a nucleus or an electron would be very small. Even if it collided, the direction of its reflection would be haphazard, depending on how the collision occurred. Is Einstein's claim not a clear evidence of misrepresentation?

Why was the simple, logical, and proven fact that light is a wave of ether, which explained everything beautifully, discarded and replaced by two contradictory and illogical theories, namely the photon theory and Einstein's version of the electromagnetic theory, which not only fell short as an explanation but also created many unsolved problems? Einstein himself admitted that the two theories are contradictory and difficult to reconcile.[218] Did all scientists

---

[216] - For further details, see Appendix G on Maxwell's theory.

[217] - Albert Einstein and Leopold Infeld, *The Evolution of Physics*, Simon and Schuster, New York, (1938), p. 99.

[218] - Ibid, p.294.

accept these contradictory theories? - or did Einstein, by misrepresentation and the power of media, impose his ideas on others?

All agree that the series of radiations such as light waves, radio waves, x-rays, heat waves etc. are waves differing only in frequency. How and where does the photon theory, which says that light is a particle, fit into this line of understanding? How did Einstein get away with the confusion he created?

At the time that Einstein presented his papers on relativity, all the experiments, without any exception, supported the existence of ether. Why did Einstein's papers, which contained absolutely no solid proof against ether, become the basis of rejecting ether?

The discovery of the radio wave was made on the basis of the existence of ether, yet the radio wave, which is a clear proof of ether, has been misinterpreted in text books and brought into line with Einstein's ideas. Why? Is it because everyone followed Einstein blindly?

Before Einstein, all scientists knew the nature of magnetic and electric forces, as it was understood that these forces arise from the effects of the flow of ether. The only single problem that had remained unsolved was how to explain the force of gravity. Einstein, by rejecting ether, not only did not solve the gravity problem, but rather, created additional problems, as his ideas could not explain any of the forces, and no one was able to generate a theory that could explain why these forces are able to act at a distance. Why, then, have these facts been ignored, and instead

Einstein's ideas, which have led nowhere, been promoted as if they were facts?

In all the books Einstein and his followers have written about relativity, their explanations and examples always arbitrarily choose a direction of motion showing an apparent contraction of length.[219] They never choose a reverse direction, which, instead of contraction shows an apparent elongation.[220] Why, despite such an obvious flaw in Einstein's idea, which also proves that it is wrong, has it been called a theory, and used against ether?

With the advent of space exploration in the 1960s, scientists discovered the earth's magnetosphere, the bow shock and the solar wind, which at once explained why Michelson and Morley had obtained no results in their search for the ether wind. The wind they were looking for exists beyond the bow shock but not inside it, which was all they could reach. The bow shock and the magnetosphere also explains that the earth drags ether. Why is it that this remarkably clear and powerful explanation which shakes the very foundation of relativity, has not been connected with Michelson's experiment?

Why has Einstein's so-called General Theory of Relativity, which solves nothing about gravity, been praised as the answer to the problems of gravity?[221]

---

[219] - See page 189.

[220] - For further details, see Appendix.

[221] - For more information, see Chapter 12. You will find that the spin of the nucleus in ether solves all gravity-related problems. A summary of Einstein's explanation is also presented, so that the reader may compare.

In 1913 Sagnac sent two simultaneously emitted light signals in two opposite directions around a closed path, and by photographic plate recorded the interference fringes where the signals met. The apparatus was placed on a turntable. He showed that when the table was rotated, it took one of the light signals a shorter time, and the other one a longer time, to reach their final meeting place than when the turntable was stationary. Why have the experiments such as the one conducted by Sagnac, which disproves Einstein's relativity, not been mentioned in any text books?

The Michelson-Gale (1925), experiment showed on a much larger scale that the velocity of light at the surface of the earth is less in the direction of the earth's rotation than in the opposite direction. In 1938, Einstein claimed that *no such difference is detected.* [222] Is this not a clear evidence of fraud?

Why have experiments disproving the photon theory[223] not been mentioned in any text books?

Professor Dingle, who also disproved the relativity theory, writes:

> *I can present the matter most briefly by saying that a proof that Einstein's special theory of relativity is*

---

[222] - See Albert Einstein and Leopold Infeld, *The Evolution of Physics*, Simon and Schuster, New York, (1938), p.179.

[223] - See Sir Edmond Whittaker, *A History of the Theories of Aether and Electricity*, New York, (1962), p. 94-95.

false has been advanced; and ignored, evaded, suppressed...[224]

G. B. Brown, Ian. McCausland and Luis Essen are examples of scientists who have tried to show the fact that criticism of Einstein's ideas has been suppressed.[225]

Petr Beckmann, a Professor Emeritus at the University of Colorado, writes:

> *Why, then, can objection to the Einstein theory be published only in the "underground" scientific press?*[226]

Wallace Kantor, who carried out important experiments in physics, writes:

> *The nominal leadership of the society of physicists is so strongly committed to a mistaken and unsupported belief in the physical validity of the Einstein special theory of relativity that explicit publication of the contradictory experimental results, in the respected physics journals, has been very effectively*

---

[224] - Herbert Dingle, *Science at the Crossroads*, Martin Brian & O'Keefe, London, (1972), p. 15.

[225] - G. Burniston Brown, a letter in *The Listener*, Vol. 88, No. 2275, 2 November 1972, p. 606.

Ian McCausland, *Einstein's Special Theory of Relativity Right or Wrong*, Toronto, (1973), p. 2.

L. Essen: "Einstein's Special Theory of Relativity", *Proceedings of the Royal Institution of Great Britain*, Vol. 45, 1972, p. 141-160.

[226] - Petr Beckmann, *Einstein Plus Two*, The Golden Press, Colorado, (1987).

suppressed. It is the primary task of this monograph to present, out in the open, this shockingly arresting evidence. No new theory is presented.

The suppression of contrary ideas by the establishment is a well chronicled historical fact that hardly needs reiteration by fresh examples of the autocratic and capricious means of its accomplishment. The suppression of specific experimental results and critiques, not in harmony with establishment beliefs, is a particularly grave situation for the very survival, much less any progress, of physical science.

The Einstein theory protagonists have been most reluctant to examine objectively any critique adverse to the theory, attributing such endeavours to a high vocal minority of crackpots, if not hopeless cranks. The easy comfort afforded by such intellectually inexpensive ridicule reveals the flavour of the situation. The ridicule has been, however, all too frequently, a case of the pot calling the kettle black. Examples of cavalier and contemptuous comment can be found in some of the very remarks made by Einstein with regard to adverse experimental results and those whom he regarded as antagonists.[227]

R. M. Santilli, a physicist at Harvard University, an 'insider' in American academic circles, came to the point where he questioned if there existed a conspiracy on

---

[227] - Wallace Kantor, *Relativistic Propagation of Light*, Coronado Press, U. S. A. (1976), p. 1.

Einstein's relativity by a few individual followers of Einstein.[228]

Why have Einstein's papers, which are full of misrepresentations and errors, been presented as the supreme achievements of the human intellect in history and none of the views of many scientists who proved otherwise ever been presented?

If Einstein's works are so great, why is access to most of them made so difficult? Why are most of them not even translated into English so that scientists can benefit from them?

What a coincidence that Hermann Minkowski, who happened to be Einstein's mathematics teacher, also became known as a world famous physicist and another genius.

What a coincidence that Einstein and Minkowski were the only two persons in the whole world who could understand the fourth dimension idea of relativity.

All this explains why impossible fictions such as, "a mother after travelling with a time machine returned to see her age is 20 but her son's age is 50", have been promoted as science. All this explains why the following fictional ideas have contaminated the realm of science: time can metamorphose into space and vice versa; empty space bends; gravity is geometry; space is empty and without ether; a body moving in empty space gains extra mass out of nothing; a moving clock will run slow because time itself is

---

[228] - Ruggero Maria Sanilli, *Ethical Probe on Einstein's Followers in the U.S.A.*, Alpha Publishing, U.S.A., (1984).

slowed by motion; a star as large as the sun can shrink into a point infinitely small and create a density infinitely large; and finally, no relative speed can exceed the speed of light, which is like saying that 200,000 km/sec + 200,000 km/sec = 300,000 km/sec, etc.

Why have a few individuals been able to get away with making the science of physics into science fiction?

All the evidence of fraud which I have found in Einstein's papers and books is only the tip of a very big iceberg. Unfortunately, due to the lengthy explanations it would take to expose it fully, only the most easily accessible have been presented here.

Chapter 33

Einstein's Ideas

The following is a general review of Einstein's ideas, followed by reasons why they are impossible. It should give you sufficient information to see how mankind has been misled.

Time Travel

The following reasons illustrate the impossibility of either going back in time or going into the future.

Consider your physical body at the present time. Your physical body is a combination of innumerable atoms gathered together from innumerably different locations. Some could have come from the remotest parts of the earth or even outer space. Through time they have assembled themselves into your body. A hundred years ago, long before you were conceived, all the atoms that presently make up your body were in different places. Suppose you could go back in time, let us say one hundred years. All the atoms of your body must also go back in time and be where they were one hundred years ago, otherwise we cannot say it is one hundred years ago. This means that by going back in time, all the particles of your body would then be disintegrated and scattered all over the earth. Nothing would be left of the form of your body to be called a human body. This suggests that the fanciful idea of going back in time is impossible. Furthermore, in every instant of time, each and every atom of the universe has a particular location and position in relation to each other. If all the positions of all the atoms are not the same, then the time is also not the same. To go back in time means that all the atoms of the universe must move back and reposition themselves in order to duplicate the exact situation as before. Now let us consider a case in which different people want to go to different times, some into different futures, some into different pasts. Given this scenario, all the atoms of the universe would be in total confusion and chaos, as each atom would not know which direction it should move in order to duplicate or recreate particular situations for a particular person. This means that each and every atom of the universe must continuously try to be in infinite locations at any instant of time in order to recreate or duplicate many particular situations for different individuals, which is obviously impossible.

Einstein and his accomplice Minkowski for almost a century misled mankind with these kinds of ideas which, while they may hold great appeal for the imagination, can in no way be considered scientific.

H. Dingle's 'Clock Paradox'

The following is how H. Dingle, a professor at the University of London, disproved Einstein's version of relativity:

> *According to the special theory of relativity, two similar clocks, A and B, which are in uniform relative motion and in which no other difference exists of which the theory takes any account, work at different rates. The situation is therefore entirely symmetrical, from which it follows that if A works faster than B, B must work faster than A. Since this is impossible, the theory must be false.*[229]

---

[229] - Herbert Dingle, *Science at the Crossroads*, Martin Brian & O'Keefe, London, (1972), p. 45.

Why has this remarkably clear refutation of Einstein's theory been ignored and not been mentioned in any text books?

Atomic Clock Experiment misrepresentation

Supporters of Einstein's version of relativity claim that an atomic clock on board an airplane would run more slowly than when it is stationary, on the ground. In 1971, Hafele and Keating conducted an experiment in which they were able to find a measurable discrepancy between the times of a stationary and a moving clock. Einstein's supporters point to this as a proof that Einstein's theory of the relativity of time is correct: the faster we move, the slower becomes the lapse of universal time. The following reasons prove that such claims have been based on misrepresentations and distortion of facts.

Firstly, Lorenz long before Einstein, on the basis of ether, had predicted that a moving clock would run more slowly than when it is stationary. The atomic clock experiment proves Lorenz's prediction and not Einstein's idea. The experiment also confirms the existence of ether from a common sense point of view, because it is understandable that a moving clock would experience a resistive pressure of ether (magnetic pressure). As the clock moves through ether, the pressure of ether slows down the parts that oscillate in the atoms of the clock. In the same way, a moving mechanical watch with its movements exposed to open air will run more slowly by the pressure of the wind; or, an electrical clock moving in the magnetic field of the earth is slowed down by the field (the current of ether). This is because the motion of a circuit in a magnetic field creates an electrical current in the circuit which may interfere with the normal current of the clock's circuit. An atomic clock is basically the same as a mechanical clock in that both systems function by the oscillation of their parts. In brief, according to Lorenz, ether is responsible for the slowing down of the clock.

On the other hand, according to Einstein and his supporters, a moving clock runs slow because time itself is slowed down. This idea is based on the speed of light. What has the speed of light to do with the time of a clock? As you can see, there is no justification for such an unreasonable and unrelated conclusion.

Furthermore, according to Einstein and his supporters, the atomic clock experiment proves that universal time itself slows down. What does the slow or fast oscillation of the parts in a mechanical or atomic clock have to do with the universal constant of time (bearing in mind that an atomic clock in principle is the same as a mechanical clock since both systems function by the oscillation of their parts)? It is very clear that relativity supporters have misrepresented the experiment in favour of Einstein's idea, and the atomic clock experiments that actually prove Lorenz' prediction and further prove ether, have been claimed as proof of Einstein's ideas. Furthermore, concerning the Hafele-Keating experiment and whether it supports Einstein's theory, Rudakov writes:

> ... *when the details of their procedure, and all assumptions on which it is based, are closely examined, it is at once evident that there cannot possibly be any connection between their results and the theory. . . . The claims advanced by relativists in support of their theory are without any foundation*[230]

---

[230] - See Clark, *Einstein, The Life and Times*, The World Publishing Co., New York, (1971), p. 148-149.

Einstein and His Fame

There is a parallel between art and science. In the world of art, you will often find that a painting which, from an artistic point of view, is not worth the canvas it was painted on and clearly shows that the person who painted it had neither talent nor skill, has a market value of millions of dollars. Why would anyone pay such a high price? The answer is: through organised marketing and thorough manipulation of the public mind by vested interest. Mock purchases and sales at very high prices create artificial market value for art. For example, auction houses can create world-wide fame for a painter through artificial purchases at very high prices, followed by a world-wide publicity campaign. At the same time, this creates an artificially high market value for the paintings.

The following is a photocopy of an article in the Toronto Star showing a photograph of a painting that was purchased by the National Gallery of Canada for 1.8 million dollars. [231]

---

[231] - *The Toronto Star*, July 16, 1993, p. A3, and July 25, 1993, p. A7.

PRETTY AS A ($1.8 MILLION) PICTURE

National Gallery officials give members of the media a preview yesterday of a new acquisition. American abstract impressionist Mark Rothko's painting, entitled No. 16, cost $1.8 million. The purchase is already controversial but officials point to earlier acquisitions that raised hackles of critics who insisted it's not art: controversy will keep the turnstiles humming.

Portrait of a major art buy

Gallery had 'test drive' of No. 16

By Henry Mietkiewicz
TORONTO STAR

OTTAWA — A little over a year ago, the National Gallery of Canada quietly embarked on what can only be described as a discreet, multi-million dollar "test drive."

Word was relayed to an agent at an exclusive New York gallery that Canada's national art museum was interested in acquiring the abstract painting by Mark Rothko, known as No. 16.

At the time, the picture — destined to provoke howls of outrage when it permanently entered the Ottawa collection earlier this month — was valued at somewhere between $3 million and $4.8 million U.S.

That was far too steep for an institution supported by the public purse. But officials at the National Gallery were eager to see first-hand whether No. 16 was really as exceptional as they'd been led to believe.

Would it be possible, they wondered, to have the 3-by-2.5-metre (10-by-8-foot) painting shipped to Ottawa and mounted for inspection?

Yes, came the reply.

And so, in June, 1992, No. 16 paid its first, unheralded visit to Canada and was installed briefly on the very wall that would later become its home.

It rested only steps away from American painter Barnett Newman's Voice Of Fire, whose apparent simplicity and hefty $1.76 million price tag sparked a fierce public outcry in 1990.

Nevertheless, the "test drive" worked with smooth efficiency.

"Seeing it there made me realize it was infinitely better than I could ever have imagined," says the gallery's chief curator, Brydon Smith, his quiet voice barely betraying his excitement.

His colleagues agreed. One by one, they threw their support behind him.

'It practically throbbed with life and vitality'

Everyone agreed there seemed to be an almost mystical quality to the way the room's other abstract artworks complemented Rothko's two roughed-edged, creamy-white rectangles, set against a glowing, orange-red background.

"It had an almost luminous quality," Smith whispers. "It seemed to pulsate. It practically throbbed with life and vitality."

But a purchase of this magnitude had to weather intense scrutiny by gallery staff, the board of trustees and the board's acquisitions committee, compounded by a prolonged period of painstaking negotiation to drive the price as low as possible.

Gallery director Shirley Thomson notes, "There are checks and balances every step of the way. Especially with a piece this costly, we engage in a full and vigorous debate at every level to ensure we're making the right moves."

All purchases are handled in such a way, she says, that, over the course of years, money and acquisitions are spread as fairly as possible through the gallery's 11 departments (Photographs, Inuit Art, Asian and Neo-Western Art, Contemporary Canadian Art, etc.).

Thomson and Smith both emphasize that the National Gallery is not depriving itself of worthwhile Canadian work, just because it eventually devoted $1.8 million (Canadian) to a single, American painting.

But the National Gallery has a special mandate, they say, of acquiring key representative pieces from other countries and other eras, so befits a leading institution of national stature and international scope.

$1.8 MILLION UNVEILING: National Gallery director Shirley Thomson and chief curator Brydon Smith show off No. 16, a painting by the late American artist Mark Rothko, inset.

Rothko still sparks controversy

By Susan Walker
TORONTO STAR

Controversy seems to cling to American painter Mark Rothko, even 23 years after his suicide.

The 67-year-old artist was found dead in his studio, his wrists slashed.

A lengthy legal dispute erupted between his heirs and his former dealer Rothko's daughter sued Marlborough Fine Arts (which stored some of the works in a Toronto warehouse) for control of his estate, later valued at $25 million (U.S.).

Whatever despair drove Rothko to take his life could never have been associated with his professional status. He was a well-established painter whose works were highly sought in the early '60s.

It was then that the women's volunteer committee at the Art Gallery of Ontario committed a large part of its funds to purchase a 1962 canvas, No. 1, White And Red, fresh from his studio.

By then, Rothko was well down the road toward the mystical, essential quality he had been seeking in his large, stained canvases. From the 1940s on, his signature became the huge, soft-edged rectangles, colored and juxtaposed to suggest a meditative, ethereal state.

A prominent member of the New York school, Rothko was born in Latvia in 1903 and came with his parents to the United States in 1913.

He was a leader among the first generation of American abstract artists, and his sense of the timeless and sublime was best achieved in his 14 panels for a Houston chapel commissioned by Dominique de Menil.

Curator Roald Nasgaard of the Art Gallery of Ontario said, "His work was important for Canadian painters like Guido Molinari, André Geuclier and Jack Bush."

The acquisitions committee, who meet four times a year to review purchases that have been completed or are being proposed, were behind the idea of buying the painting.

Five members of the committee are drawn from the board of trustees, while the remaining six are community leaders or art experts, such as Phyllis Lambert (founder and director of Montreal's Canadian Centre for Architecture), Vancouver art critic and historian John O'Brian, and Montreal ophthalmologist Dr. Sean Boiler (former board chairperson of the National Museums of Canada).

The chief stumbling block, of course, was money.

And so, in the spring of 1992, negotiations began: National Gallery officials gave instructions to Smith, who conferred by telephone with the Pace's Douglas Baxter, who contacted the Rothko estate, who sent word back through Baxter and then through Smith to the National Gallery.

By June of 1992, the asking price was down to about $3 million.

It was at this critical juncture that No. 16 made its quiet trip to the National Gallery. As the summer drew to a close, Smith had managed to pare the figure to $2.5 million (U.S.). He believes this was possible because Rothko's estate knew the painting would not be acquired by a private speculator, but, rather, it would be exhibited to and enjoyed by the public among similar artworks in a modern institution of national stature.

But the figure still wasn't low enough. According to Hornstein, the acquisitions committee recommended at its next quarterly meeting in September, 1992, that the painting be purchased only if Smith could get it for $1.8 million (U.S.).

Shortly thereafter, the matter was presented to the board of trustees, whose 14 members are appointed for three-year terms by the communications minister. They include Jean Claude Delorme, former board chairperson of a $41 billion Quebec pension fund, painters Alex Colville and Ken Denby, art collector and theatrical producer David Mirvish.

No politicians sit on the board of trustees, but its members answer to the government's standing committee on communications and culture.

Armed with detailed, technical reports about the condition and intrinsic value of No. 16, the board adopted the recommendation of the acquisitions committee and instructed Smith to shoot for $1.5 million (U.S.).

"It was $1.8 million or nothing," Hornstein says.

Not long after New Year's, Smith finally got the news: Rothko's estate had agreed to the deal.

No. 16 was finally installed on July 13 and unveiled three days later, to a flurry of criticism and no small degree of ridicule.

Standing before the painting, Smith pauses. "Whatever effort it took to get No. 16," he says quietly, "it was worth it."

"We want to present a panorama of art," says Michal Hornstein, chairperson of the acquisitions committee. "Sometimes this means it makes more sense for us to spend $1 million on one major work than on another 50 paintings at $20,000 each."

For 25 years, the notion of buying No. 16 was not an issue anywhere in the world, because no one knew where the painting was.

Completed in 1957, No. 16 became one of Rothko's most popular pieces since, like abstract work by other New York artists of the period, it skillfully used the tone, texture and expressive color to ignite the viewer's emotions.

So provocative was No. 16 that it was included in a European tour of Rothko's work. And, on its return to New York, the artist himself placed it on an inventory list and had it crated and stored in a midtown warehouse.

Then, somehow, it got lost. Rothko, who died in 1970, never saw his painting again.

Only in 1988, when the warehouse was slated to close, did No. 16 finally resurface during a detailed inventory. The canvas promptly returned to Rothko's estate — represented by the Pace Gallery of New York — and placed on the open market.

Two years passed, and No. 16 languished unsold, possibly because of its steep asking price of $4.8 million (U.S.).

Still, Douglas Baxter, vice-president of the Pace Gallery, says he received indications of interest from at least two American museums and a "serious inquiry from a national gallery in Europe."

At the same time, Baxter continued to send transparencies of No. 16 to anyone believed to be in the market for a large Rothko. Which is how Brydon Smith came into the picture.

"After the acquisition of Voice Of Fire," Smith recalls, "it seemed obvious to me and my colleagues that a Rothko would add something unique to the National Gallery."

Back in the mid-'80s, while serving as a curatorial assistant at the Art Gallery of Ontario, Smith received what he calls his "first intimate introduction" to No. 1 White And Red, the only other large Rothko in a public Canadian collection.

Intrigued by the transparency of No. 16 that he'd received in late 1990, Smith dropped into the Pace Gallery in the spring of 1991, during one of his regular trips to New York.

From the start, the 11 members of

Looking carefully at the painting, one sees that it is mainly blank, and devoid of anything that could be called art. In the same newspaper[232] a few days prior, several copies of the same painting were shown that had been painted by children in a few minutes. The copies as shown in the paper were so exact as to defy anyone to tell the difference between the children's work and the original. The question that came to my mind was: how many other paintings of a similar nature have been purchased with public money, paintings with less money and no publicity?[233] How many similar paintings have been donated to museums, in return for which the donor received a receipt on which he or she could claim a tax exemption? The total amount could be billions of dollars. The fact is that some museums have amassed tens of thousands of these kinds of paintings in their storage.[234] In addition to these paintings, how many were sold to different individuals who fell into the trap of purchasing them in the hope of one day selling them for a profit?

How does the general public come to accept a certain painting as a great work of art when their own eyes tell them it is not? How many times, in a museum or art gallery, have you seen such a painting yet nevertheless said to yourself, "*Perhaps I do not understand it, perhaps there is something in it that I cannot see*"? After hearing repeatedly about the painting and its price, you may then have said to yourself that the painting must be a great artistic work. The

---

[232] - *The Toronto Star*, July 19, 1993.

[233] - Also, paintings with artificially high market values are donated to museums and then the donor receives a receipt entitling him or her to an equivalent tax exemption.

[234] - A television documentry revealed that museums have been stacking tens of thousands of paintings in their storage area.

fame of the painter, the price of the painting and its place in the museum create in your mind a new opinion for the painting, an opinion which is artificial and the product of manipulation of the mind.

In the world of science, a similar parallel exists, with the difference that the process is not as visible as in the arts. When you see a painting, everything about the painting itself is in front of you. The merit of a scientific paper, however, is not as visible as a painting. You cannot see everything at once, especially if the paper is written in such a way that everything is ambiguous and mixed with complicated mathematical language. For this reason, a scientific paper is more difficult to judge at first glance.

On November 10, 1979, at a scientific conference where a large number of scientists had gathered in honour of the centenary of Einstein's birth, many papers describing and reviewing the works of Einstein were read. At that conference, Einstein's main contributions to the scientific world were identified as (1) the special theory of relativity; (2) the wave and particle theory; and (3) the general theory of relativity.[235]

In the previous chapter, it was proven that the special theory of relativity does not belong to Einstein, yet he, without any justification, received credit for it. It was also shown that he misrepresented and distorted the facts by suggesting that a false appearance to an observer is not false but true. He downgraded science from truth to illusion. In order to accomplish his purpose, he attempted to disprove the existence of ether. He did this by claiming that relativity has nothing to do with the effects of ether, but rather with the speed of light. He created the false idea that relativity

---

[235] - *Van Nostrand's Scientific Encyclopedia*, Sixth ed., Considine Editor, Van Nostrand Reinhold Company, New York, London, p. 2427.

disproves ether. By creating false information about ether, he was able to make people believe that ether had already been disproved. Through these misrepresentations, Einstein was able to remove ether from science and effectively destroy what had taken many scientists centuries to accomplish.

The second of Einstein's so-called 'contributions' was the particle and wave duality idea. It was mentioned before that Einstein had to disprove ether in order to prove his relativity theory. By removing ether from physics, he then had to produce an explanation for light without ether, because up to that time all scientists regarded light to be a wave of ether. For this reason he tried to prove that light is a particle which moves through empty space without ether. Since the wave nature of light was a proven fact which could not be disputed, Einstein then added to the particle theory the electromagnetic wave theory, allegedly taken from Maxwell. However, in removing ether, he misrepresented Maxwell's theory, for ether is the very foundation of Maxwell's theory. It must be noted that, according to Maxwell's theory, light is created by electromagnetic disturbances in ether and is propagated in space through the medium of ether. Einstein misrepresented the theory to make it appear as if light is an electromagnetic wave that moves without ether. In short, the wave aspect of the theory is based on a misrepresentation and distortion of Maxwell's theory.[236]

The particle theory (the photon theory) which is also regarded as one of Einstein's important contributions to science, is another misrepresentation and distortion of facts. Despite the fact that the photon theory has been proven to be incorrect, contrary to the laws of physics and scientifically

---

[236] - For detail, see Appendix, F, "Evidence of Fraud: Einstein's Misrepresentation of Maxwell's Theory."

unsubstantiated, yet it has been made a part of science. Even the alleged 'proof' that Einstein invented does not support it.[237] The accommodation of Einstein's so-called particle theory creates the absurd and impossible particle-wave duality theory in which light is in all cases a wave, except in the case that Einstein happened to give an opinion on, in which case light is a particle. Here is what Einstein himself claimed:

> *One of the most fundamental questions raised by recent advances in science is how to reconcile the two contradictory views of matter and wave. It is one of those fundamental difficulties which, once formulated, must lead, in the long run, to scientific progress.*[238]

Since Einstein's particle and wave theory, one of his so-called 'most important contributions', has lead science into contradiction, then, on what basis is he considered to be the greatest scientist of all times?

In brief, Einstein's alleged 'advance' was that he destroyed centuries of progress in the understanding of light. Collier Encyclopaedia writes:

> *...the theory of relativity did not answer the fundamental question of how light is propagated and left things essentially as in the days of Young and Fresnel.*[239]

---

[237] - For detail, see the photon theory.

[238] - Albert Einstein and Leopold Infeld, *The Evolution of Physics*, Simon and Schuster, New York, (1938), p.294.

[239] - *Collier's Encyclopedia*, (1987), v. 14, P. 627.

Finally comes Einstein's paper about gravity, his "General Theory of Relativity". First of all, it has been shown that the theory, despite its name, has nothing to justify itself to be called General Relativity, as it has nothing to do with relativity. In Chapter 12, we reviewed the general theory of relativity and demonstrated why Einstein's paper contributes nothing to science. According to Einstein's theory, everything is attracted toward the centre of the earth because time bends and space curves. What has a force that causes the fall of objects to the ground to do with the curvature of time and space? How could one accept that a physical body has weight because space curves and time bends? How could anyone accept the idea that two cubes located side by side have different weights because time and space are not flat (bearing in mind that as you are looking at the cubes, the time and the amount of space for both cubes is exactly the same)? Have you ever questioned: how can empty space curve? How can time, which only goes in one direction, reverse and go backward? These ideas may appeal to the imagination, but cannot be justified scientifically.

Einstein uses an example in which someone is in an elevator which is being accelerated by a hypothetical force. Einstein then argues that because the person in the elevator cannot differentiate between gravity and inertia, then these two phenomena are equivalent. What does this tell us about gravity or inertia? Einstein has made no advancement to our understanding of the forces involved. He has explained neither inertia nor gravity; yet people believe that Einstein solved the riddle of gravity. Einstein's use of the idea of "equivalent" is similar to saying that since the qualities of this force are equivalent to those of another force, we have defined the nature of the force. In other words, according to Einstein, if 100 horse power of energy produced by Source 'A' is equivalent to 100 horse power of energy from Source

'B' we have defined the nature of the sources, how the energy is produced and how it makes itself felt at a distance.

What does this kind of equivalence have to do with the nature of the energy and how it is produced? Is the energy being produced by sunlight, electricity, air, a car, a horse, a jet propulsion, or by human beings? Using complicated mathematical manipulation and ideas such as space curves and the fourth dimension borrowed from Minkowski, by all manner of hypothetical assumptions and complicated equations, formulas and ambiguous interpretations, Einstein has created enough distractions to make scientists reading the paper so confused that they lose sight of how meaningless the ideas are as actual contributions to scientific knowledge.

In brief, Einstein is credited with having found some answers about gravity, but in reality he did nothing to illumine scientific understanding of the subject. He failed to answer this simple question: - why is everything attracted toward the centre of the earth? (See chap. 12 for the ether explanation of gravity). Despite this fact, the paper has been praised as one of the greatest achievements in the history of science. Einstein claims his theory solves one of the problems about the motion of planets. Yet, after reading what experts had written about it, I realised that the theory not only solves nothing but creates more problems.[240]

George de Bothezat, one of the respected scientists of the time, whose lectures Einstein and many famous scientists attended, wrote about Einstein's theory of relativity:

> *It certainly is not to the credit of our time that the greatest blunder ever committed in science was*

---

[240] - See Charles Lane Poor, *Gravitation Versus Relativity*, (1922), G. P. Putnam's Sons, New York, London, p. 19-21.

so lightly accepted as a great verity and was permitted to achieve such fame, although those who have mostly glorified this assumed great discovery really knew next to nothing about it.[241]

About Einstein and his works, Bryan G. Wallace in his book *The Farce of Physics* writes:

Einstein's theories and his status as a scientist are at the core of the problem of modern physics being an elaborate farce, ...[242]

---

[241] - George de Bothezat, *Back to Newton: a Challenge to Einstein's Theory of Relativity*, G. E. Stechert & Co., New York, ((1936), p. 151.

[242] - Bryan G. Wallace, *The Farce of Physics*, St. Petersburg, Fl., (11993), Fax: (813) 864-8382, Email Wallace@eckerd.edu.

Chapter 34

Humanity's Loss: The Ether Propulsion System

I believe if Einstein had not misled mankind to reject ether, by now many wonderful spaceships would have been invented that would have enabled mankind to travel to other planets, spaceships which would function with the force of ether.

A spaceship moving by the force of ether could travel at speeds comparable to the speed of light. A jet or airplane that moves by the force of gas or air can only travel at speeds comparable to the speed of sound. A spaceship utilising the force of ether would enable us to travel to different planets. By now we could have explored many planets. Travelling to different planets or simply to different continents on earth would be easy and very rapid.

The following are some suggestions about how such a spaceship could be built. From the way that an airplane's propeller takes in air from one direction and then pushes it away in another direction and as a result creates propulsion to fly, the thought came to my mind to invent a propulsion system, a device, that could do the same thing with ether. After a few years of work, I designed a system that could possibly be used for air and space transportation. In 1990, a patent was granted for my invention; however, no organisation was found ready to make a prototype which could prove or disprove its feasibility. I am anxious to see if the specific design does, in fact, work. Although the principle is correct, it requires testing and refining. There are many obstacles to overcome, as is normally the case in such sophisticated inventions.

The basic principle of this system is very simple. If one end of a conducting wire ("A") is connected to a source of very high voltage electricity, and the other end of the wire ("emitter") is made very sharp, we find that at the emitter there is a force generated which tends to move the wire. The reason for the force is this: in the conducting wire, the high voltage, which is in reality a high pressure of ether, allows for some of the ether from the emitter to escape into space. Since space is filled with ether, the ether that escapes exerts a pressure on the conductor. As this high pressure ether leaves the emitter, it carries along some electrons that enhance the pressure. The higher the voltage, the greater the pressure. In a flat panel, millions of tiny emitters are arranged side by side and embedded within the panel, which is made of insulating material. All the emitters are parallel to each other, and connected to an electromagnetic power source. As each and every emitter generates pressure, the total combination of the force generated by all the emitters amounts to a significant propulsive force that can lift or move an object in space. Fig. 1 shows a panel with a large number of emitters. These emitters are embedded within the panel, and normally are not visible. Since the emitters cannot touch one another and should be separated by insulating material, there is a limit to the number of emitters that could be used in a given panel.

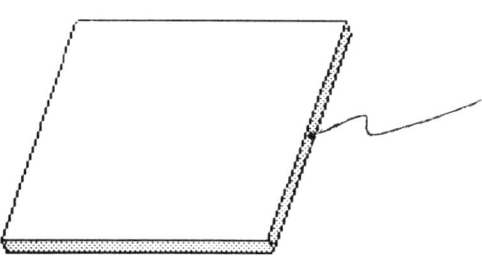

Fig. 1

For this reason, with a given power supply, each panel generates a limited amount of force. Therefore, a spaceship using these panels as its driving force must have a large exterior surface area, such as the design we associate with a typical flying saucer, in order to have the required number of panels.

A typical panel of 1 square meter in size accommodates as many as 250,000 emitters, each emitter being 2 millimeters apart from the adjacent emitter. Each emitters creates a force that can lift a weight of .4 grams. Each panel can lift a weight of up to 100 kilograms. A typical spaceship with a 20-meter radius can have a winged area of 1256 square meters. If this area is completely covered with panels, all the panels together create a force that can lift a weight of 125.5 tonnes, a weight that is still greater than the weight of a fully loaded 747 jumbo jet.

With a special change in the way the emitters are covered by insulating material, the propulsive force can be changed into a force of suction or a reverse propulsive force, which takes in ether and attracts electrons rather than losing them. These panels can also be used to lift or move objects in space. The combination of propulsive and reverse propulsive panels could be used to reinforce each other and at the same time avoid excessive charging or recharging of the system. Furthermore, each panel could be supplied with a charge-collector screen system by using a secondary circuit that further prevents the system from losing electrons.

Two more methods to prevent the charging and discharging of the panels would be: first; to find a conductor that conducts ether without electrons; and second,[243] instead

---

[243] - The methods that have been suggested here are not the only possibilities. The whole system could be designed quite differently; instead of metal wires that carry electrons, non-conducting wires made

of supplying the panels with ultra high voltage, to connect the emitters to a high-frequency electromagnetic power supply with sufficient voltage to avoid interference between emitters and have radiating emitters. Since it is a known fact that radiation can apply pressure, the radiating panels, by high voltage radiation, become propulsive panels.

Fig. 2 shows the shape of a typical spaceship, which must be round to maximize wing area. Its exterior surface, top and bottom, is covered by panels, which are electrically connected to an electromagnetic power generator located inside the spaceship. The amount of electricity to the panels is controlled by a pilot inside the spaceship.

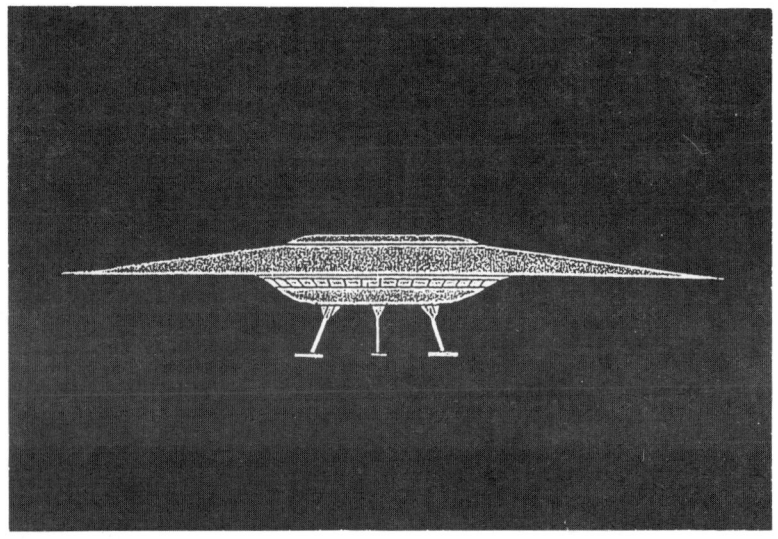

---

of special materials may be used to carry the ether, without any electrons, to the points.

ADVANTAGES OF THIS SYSTEM:

1 - In a jet or rocket system, the area that generates the propulsion compared to the entire exterior body of the jet or rocket is very small, and most of the exterior surface creates friction and resistance against the air. This means that a large amount of energy is wasted in overcoming friction and resistance. This is especially true at high speeds, for the higher the speed, the greater the friction and air resistance. The ether propulsion system eliminates all friction and resistance, because the entire surface of the spaceship generates propulsion. There is no area left for resistance or friction.

2 - Since the entire exterior surface of the spaceship generates propulsion, the force of inertia (which is the resistance of ether to acceleration) is eliminated. This means that a spaceship using ether propulsion can break one of the known laws of physics, that is, it can pick up speed almost instantaneously. It can stop almost instantaneously. It can reverse its direction by 180 degrees almost instantaneously. This means also that the system can reach speeds close to the speed of light. Unlike rocket systems, the ether propulsion system can easily continue to generate propulsion for a very long time.

3 - Unlike most rocket systems that can be used only once, ether propulsion can be used many times.

4 - A jet or rocket system uses enormous amounts of fuel for a given distance. For example, to send a satellite above the atmosphere, a rocket uses hundreds of tons of fuel; a 747 jet uses over 20,000 litres of gasoline to cross the Atlantic

Ocean. On the other hand, the ether propulsion system uses none, if a battery or nuclear power generator is used to supply the needed electricity.

5 - The ether propulsion system is noise free, far less pollutive, and can travel in both air and in outer space.

6 - An ether spacecraft can lift vertically without the need for long runways.

7 - The system is safe and in case of electrical failure, can glide and land safely, because of its large wing area.

This invention with all its characteristics is amazingly similar to descriptions of flying saucers that many have claimed to have seen and experienced. Could it be that these claims are really true? The possibility is great. There are many reasons to believe that besides our civilisation, there must exist many other civilisations, some of them far more advanced than ours.

Conclusion

This book, which is the result of years of research and investigation, introduces discoveries which undoubtedly will change fundamental understandings of the physical universe, let alone the science of physics. How I arrived at these discoveries is another important story by itself.

In this book you were introduced to a series of important discoveries such as that of the samareh, a subatomic particle that revolves around an electron. The discovery of the samareh solves all problems related to electrons, magnetism and electricity.

But the fundamental verity which this book strives to re-establish above all is the existence of ether, a fact which Einstein not only rejected but turned the whole world against. We know that all the infinite variety of physical existence that we can see and touch is composed of a hundred or so elements such as oxygen, hydrogen, iron, etc. We know also that these hundred or so elements are made from only several ingredients, namely the proton, electron, positron, etc. Evidence tells us that these several ingredients must be made of only one ingredient which we call the ether particle. Just as the beginning of all numbers is one, so are all the diversities of creation made from one ingredient, namely ether. It is here proposed that the ether particle is the building block of all matter. The existence of ether as an invisible substance in space and as the building block of all that exists physically, can be compared to water molecules which can exist in different forms, such as invisible humidity in air, as water in the sea, or as a solid like ice or snow on the ground.

The scientists of the 19th century such as Faraday, Young and Fresnel, Lord Kelvin, Maxwell, Hertz and many others, conducted a series of experiments that led them to conclude, with absolute conviction, that space is filled with ether. They arrived at the understanding that ether particles must exist everywhere: within and outside atoms, in air, in outer space and even in all man-made vacuums. They concluded that light is from the wave and vibration of ether just as sound is from the wave and vibration of air. The understanding that ether exists led Hertz to the discovery of the radio wave.

With this understanding one realizes that due to the fact that the space around us is filled with both air and ether, many disturbances, such as the impact of a sledgehammer on a hard stone, affect both air and ether and thus generate both sound and light. That is why natural lightning, sparks, explosions – nuclear or non-nuclear, etc. generate both light and sound.

Moreover, one also realizes that the light and heat of a nuclear explosion results only from the disturbances and friction that sudden vaporisation and expansion of volume create in ether. Since both light and sound are waves, all the laws such as reflection, deflection, diffraction, refraction, the Doppler effect, interference, etc., for both sound and light are exactly the same. All the radiations, namely gamma rays, X-rays, visible light, infrared, heat waves, radio waves, etc., are simply the wave and vibration of ether, differing only in frequency. All radiations emanating from an atom are generated by disturbances that the motions of subatomic particles create in ether. Hence, gamma rays are generated by the fast spin and vibration of the nucleus in ether. Heat is generated by the fast orbital motion of electrons around the nucleus. Light is generated by the fast spiral motion of the samareh around an electron. It was shown that the fast motion of an electron or an ion in ether creates disturbances or friction in ether which in turn generate light which we call sparks or lightning. The light of comets is exactly the same phenomenon. The fast motion of a comet in ether generates friction and light. The tail of a comet is similar to the wake of a moving motor boat in water. The rings around the planets are the same phenomenon: the fast motion of a moon around a planet creates friction and tail waves in ether, which we see as a ring around the planet. The light of the sun is also from disturbances in ether. Contrary to what physicists currently believe, the light of the sun is not from unexplainable and exhaustible nuclear reactions supposedly occurring deep in the centre of the sun, but rather from violent surface disturbances that affect ether. Furthermore, the sun is not composed solely of gases, but its surface is covered by an ocean of hot and radioactive liquid that is continuously ionising the atmosphere and creating storms with such intensity and power that the ether is agitated into extreme excitation and vibration. These vibrations of ether reach us as the light and heat of the sun which is, contrary to current belief, an inexhaustible source of energy.

It was explained how the magnetic force is created from the current of ether. It was reasoned that the flow of ether has some characteristics similar to the flow of air. As the flow of air can create forces of pressure and suction, similarly the flow of ether creates the magnetic forces of attraction and repulsion. Unlike the pressure of air which can be felt on the surface of our skin, ether pressure, (the magnetic force) can not be felt on the surface of the skin because ether particles are so small that they pass through our bodies without exerting noticeable pressure.

The understanding that all space is filled with ether led 19 th century scientists to many predictions, such as the radio wave, which were subsequently proven to be true. Based on the understanding that a body moving in ether must encounter a resistive pressure of ether, Lorenz was able to develop his theory of relativity and its formulas that up to this date scientists use. Just as a body moving in air encounters the resistive pressure of air that compresses the body, a body moving in ether encounters a resistive pressure of ether and as a result, contracts in the direction of motion. The mass increase due to the motion is because the body not only encounters the resistive pressure of ether, but it also drags some of the compressed ether and as a result, the body behaves as if its mass is increased. Inertia is a force created by the resistance of ether to acceleration or deceleration just as water resists acceleration and deceleration.

One of the important discoveries in this book is that the nuclei of atoms spin, and that their spin in ether creates the force of gravity. The spin of the nucleus in ether solves all the problems about gravitational force with such simplicity that it amazes the mind. Furthermore, the spin of the nucleus explains the limited size and stability of an atom. It also explains why the nuclei of some atoms generate gamma rays. It was shown that alpha particles get their speed from the spin of the nucleus. The spin enables us to calculate the size of the nucleus. If one compares the explanations about the cause of gravity that have been provided in this book with Einstein's, one will find that on the one hand, all the explanations in this book are based on simple laws of nature and can be proven by simple experiment. On the other hand, upon reading Einstein's paper from the beginning to the end, one will find oneself bombarded by complicated mathematical equations and formulas derived from assumptions that have nothing to do with either the laws of nature or our physical experience. Interpretations and unrelated conclusions will be found that have nothing to do with reality.

All three forces of nature, namely the electric, magnetic and gravitational forces, result from the flow of ether. The spin of an electron in ether creates what we call the electric field.

The solar wind is created by the spin of the sun in ether and, contrary to what scientist believe, nothing leaves the sun's atmosphere.

It was proven that all planets and comets drag an atmosphere of ether. The earth also drags ether as it orbits the sun. With the advent of rockets, scientists have been able to conduct experiments beyond the air atmosphere, deep into space where they have discovered the earth's magnetosphere, the bow shock and beyond it the solar wind, which at once explain why the Michelson and Morley experiment, even at heights of a mountain, obtained nil results. Michelson and Morley's experiment, the phenomenon of stellar aberration, the discovery of the earth's magnetosphere, a comet's coma and other results all confirm the earth's drag of ether.

We then investigated why scientists abandoned the concept of ether at the beginning of this century, and with it the most fundamental pillars of science, such as universal time and space. It was illustrated that the abandonment of ether was not based on any scientific evidence at all but instead was due to misrepresentation and distortion of facts. Mankind has been misled for almost a century. The evidence presented is so powerful as to be eye-opening.

Einstein, a patent officer, with his 1905 relativity paper, and Minkowski with his fourth dimension idea, shrouded science in unfathomable mystery. The understanding that ether exists was replaced by ideas of empty space and the fourth dimension. Einstein changed the course of science and transformed physics into fiction and confusion. The fanciful idea that one day man can go into the future or the past blinded the minds of the masses to such a degree that no scientist dared challenge Einstein. The fact that it was so widely publicized that in the whole world there were only two persons who could understand Einstein's relativity paper (himself and Minkowski) shows how blindly and mischievously the paper was glorified as one of the greatest scientific works ever produced, and Einstein as the greatest scientist mankind had ever seen.

As noted earlier, after years of investigation of Einstein's works, I have discovered that all of his ideas actually belonged to other scientists, whose works he disguised through mathematical distortion. In the world of art, if some one steals a work of art and puts his own signature on it, such an act is called "fraud". In the world of science, Einstein took the results of a paper, which had taken someone years of research to write, and, by fabricating assumptions and adding convoluted mathematical steps backward, reached the same result. Everyone reading his papers thought that he had independently arrived at conclusions that happened to belong to another scientist. Even those who realized what was happening thought they had uncovered an isolated case. No one suspected that the same problem existed in all of Einstein's papers.

These facts led me to raise the question that Einstein's papers could amount to acts of fraud. Further confirmation of this suspicion came as I uncovered a trail of misrepresentation and false information, beginning with his fabricated evidence against the existence of ether. The misrepresentations of Michelson's experiment; the photoelectric effect formula which led to the acceptance of Einstein's contradictory particle and wave theory of light; Hertz's discovery of radio waves; Maxwell's electromagnetic theory of light; Fizeau's experiment; and many, many other misrepresentations one and all are only part of the evidence. The fact that Einstein has been portrayed as the greatest scientist mankind has ever seen is another evidence of misrepresentation. Bear in mind that no invention of any kind has ever come from any of Einstein's ideas. He has contributed nothing useful to science. The fallacy that Einstein's letter started the Manhattan project, the lauding of Einstein's poor and unscientific papers as the greatest and most praiseworthy scientific works, one and all fit the same pattern. Although many scientists have realized that there have been various misrepresentations in Einstein's works and many have even written books and articles about them, no one has seen the whole picture at once. By putting all the evidence together, a true picture emerges that exposes the reality of Einstein's works and changes all of our notions about Einstein. I realize now that the damage he has caused to the world of science is enormous. Tens of thousands of books have been written on Einstein's works. Many trillions of hours have been totally wasted by scientists and individuals on Einstein's ideas. Trillions of dollars have been wasted and continue to be spent on these ideas, all because Einstein, by his actions, has highjacked science, misled mankind and considerably retarded scientific progress. Otherwise, without doubt, mankind would have advanced to such a degree that by now interplanetary travel would have been part of our lives.

It must be pointed out that today there are many who will ignore, dismiss and even attack the clear evidence and discoveries presented in this book and cling tenaciously to Einstein's ideas at any cost. Unfortunately, a few of them may hold high academic positions and may feel that this book threatens their established positions. An example of this kind of mentality is shown in the following quotation given by a former president of The American Physical Society, in his retirement address.

> *Just suppose, even though it is probably a logical impossibility, some smart aleck came up with a simple, self evident, closed theory of everything. I and so many others have had a perfectly wonderful life pursuing the will-o'-the-wisp of unification. I have dreamt of my children, their children and their children's children all having this same beautiful experience.*
>
> *All that would end.*
>
> *APS membership would drop precipitously. Fellow members, could we afford this catastrophe? We must prepare a crisis management plan for this eventuality, however remote. First we must voice a hearty denial. Then we should ostracise the culprit and hold up for years any publication by the use of our well-practiced referees.*[244]

Although this quotation on the surface appears to be no more than humour, yet there are some who think exactly in this way, and the reader should keep this in mind.

When comparing ideas in this book with those of Einstein's, you have a choice, either to accept what is reasonable and in accordance with the laws of nature, or fall for the artificial fame of Einstein and accept his nonsense.

It is hoped that what you have read in this book has given you a clear perspective by which you may judge. It is hoped that by now you have agreed with the information that has been presented to you, and whether you are a teacher, a scientist, a student, or whatever your position in society, you will do your part by lending your voice in its

---

[244] - R. R. Wilson, *Physics Today*, 35(6), (1986), p. 30.

support. There is a great deal of further evidence to substantiate the findings in this book, which will be left for others to discover. If you do nothing, your children, your children's children, generation after generation will remain in darkness. Remember that if you do not investigate the truth, who is there to do it? With your help, together, we can accomplish that which shall leave the mark of victory for the generations to come.

APPENDIX A

Laws of Reflection of Light and Sound Waves

The laws for the reflection of sound are the same as for those of light and can be demonstrated by a simple experiment that can be found in many old physics texts. For example, a source of sound, such as a ticking watch or a high-pitched whistle, is placed at A (Fig. 1). At the end of a two- or three-foot long smooth metal tube AB, a reflector or plate reflects the sound that travels along AB, while a second tube, CD, provided at its end with a rubber tube leading to the listening ear, or some other detector of sound, is then adjusted by varying the angle it makes with E until the sound received along the tube is at a maximum.

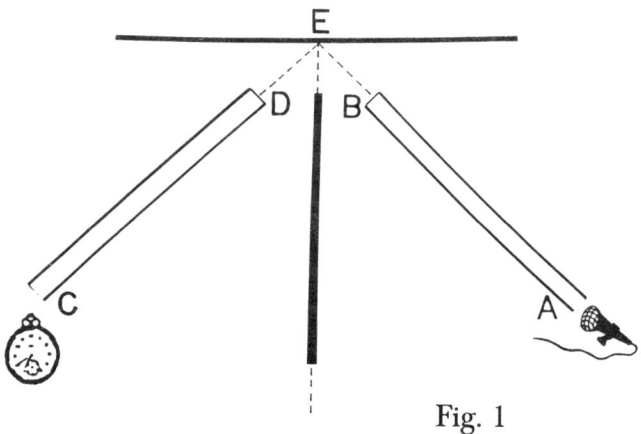

Fig. 1

It is found that the axes of the two tubes make equal angles with the normal to the reflector E. From this experiment and others of a similar nature, it is found that the angle of reflection is equal to the angle of incidence. By similar experiments using light instead of sound, it has been found that the angle of reflection of light is also the same as the angle of incident light.[245]

The laws of the reflection of sound and light from curved surfaces can also be illustrated using two parabolic mirrors twenty or thirty meters apart (Fig. 2), and by placing a source of sound such as a watch at the focus of one of the mirrors, A.

Fig. 2

The sound will be reflected as a parallel beam due to the curvature of the mirror. A second similar mirror at B collects these waves at the focus, and a funnel with a tube leading to the ear makes it possible for a listener to hear the ticking of the watch, which would be inaudible elsewhere in the vicinity of the mirror. Sound sensitivity is due to the fact that sound energy falling upon the large area of the mirror is reflected onto the small area of the microphone. Similar experiments found in physics books can be conducted with light. In the case of light, large parabolic mirrors can be used to form a sharp image of stars.

---

[245] - C.E. Mendenhall, A.S. Eve, *College Physics*, D.H. Heath and Co., Boston 1944, p. 191.

APPENDIX B

The Refraction of Light and Sound Waves

When a light or sound wave passes from one medium to another of different density or elasticity, its velocity is changed, and if the direction of propagation is altered, the wave is said to be refracted. The laws for the refraction of sound are the same as for those of light. Thus, a rubber bag in the shape of a lens when filled with a heavy gas such as carbon dioxide will concentrate the sound of a watch or a shrill whistle to a point, just as a magnifying lens of glass will concentrate the light of the sun to a point.[246]

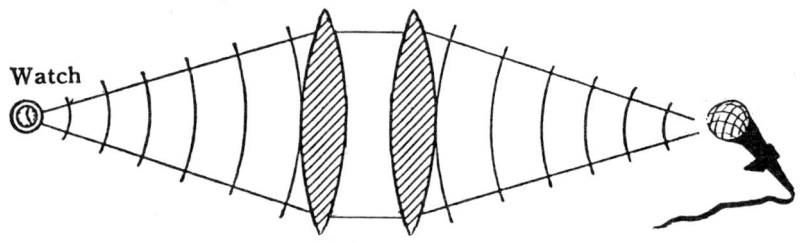

Fig. 1

There are innumerable other phenomena related to light and sound that are very similar. These examples show clearly that light is the wave of a medium, as sound is the wave of a medium.

---

[246] - C. E. Mendenhall, A. S. Eye and D.A. Kets, *College Physics*, p. 190.

APPENDIX C

The Laws of Interference of Light and Sound Waves

The laws for the interference of light are the same as for those of water or sound waves. Interference of light is one of the most clear proofs of the wave nature of light. Scientists generally agree that the photon theory (the corpuscular or particle theory of light) has failed completely to account for the interference of light. Einstein and the supporters of the photon theory could not dispute the fact that interference is the property of waves only and not that of particles, confirming that light is a wave and not a particle.

In 1800, at a time when the prevalent theory of light was that of Isaac Newton's corpuscular theory, which explained light as a particle, Thomas Young, challenging the theory, proposed that light, like sound, must consist of waves. On this basis he predicted the interference of light similar to sound or water waves. In 1803 he was able to prove his prediction by producing a series of interference bands on a screen. Detailed studies of interference are available in any college physics text or related encyclopaedia.

APPENDIX D

The Independence of the Speed of a Wave from its Source

The law for the independence of the speed of light from its source is the same as for that of sound or any other wave. This law also indicates that light is a wave and not a particle.

When a disturbance is created in a medium and a wave is formed, the medium carries the wave with a speed that depends on the nature and state of the medium. This is one of the fundamental characteristics of any wave. The speed does not bring into account whether the source is moving or stationary, advancing or retreating. The speed of a wave depends on how fast the medium transmits the wave from one place to another.

For example, the sound of a car horn travels at 1200 kilometres per hour in all directions, regardless of the speed or direction of the car. Another example is that of water waves generated by the motion of a ship. The speed or direction of the ship has nothing to do with how fast the wave travels in water. A further example is that of the sound of musical instruments travelling with a constant speed in air, which does not bring into account how fast the strings vibrate or how fast the musical instruments move. The speed of the sound depends solely on the nature of air, which is the medium carrying the sound waves in all directions.

James Bradley discovered the aberration of light in 1725. He discovered that the apparent position of the stars is not the same as their actual position. He concluded that the speed of light and the velocity of the earth are responsible for the difference between apparent and actual positions. Since all the stars show the same displacement, it follows that the velocity of light is independent of that of its source. This can easily be explained if light is a series of waves in stationary ether, but it is difficult to determine how corpuscles (photons) emitted by a moving body cannot be influenced by the speed or direction of the moving body. The discovery of aberration by Bradley confirms that light is a wave and not a particle.

The independence of the speed of a wave from the speed and direction of its source is solely the property of waves and not particles. For the speed of particles, unlike waves, are affected by the speed of their source. For example, the speed of a ball thrown by a man depends on how fast the man moves his hand. The speed of a bullet depends on the speed or direction of the motion of the gun. However, since one of the basic characteristics of light is that its speed is independent of the direction or speed of its source, this proves that light is a wave. For example, the speed of the light of a star reaching the earth has nothing to do with how fast the star is moving or in what direction the star is moving. The light of a star travels with a speed of 300,000 kilometres per second in space.

In previous chapters, it was shown that Einstein, in his relativity paper, without giving any reason or evidence, assumed that light is a particle and that the particle has the characteristics which belong exclusively to waves. He assumed that light always travels with a constant speed, exactly like a wave, and that the particle's speed is independent of its source.

It must be pointed out that the frequency of waves does depend on the state of motion of their source. So, the speed or direction of the source can affect the frequency of the wave it emits. This effect is called the Doppler effect which will be discussed in the next section.

APPENDIX E

Laws of Doppler Effect for Light and Sound Waves

The law of the Doppler effect for light waves is the same as that of sound or any other wave including water waves. You may have noticed that while a fast train is whistling and approaching you, its sound has a higher frequency (higher pitch) than when it is moving away. This is because in the case of an advancing train, the speed of the train increases the frequency of sound in your direction, whereas with a retreating train, the speed of the train decreases the frequency of sound. This phenomenon is called the Doppler effect. Frequency is also affected by the direction or speed of the receiver. In other words, the state of motion of the receiver can also create the Doppler effect. But here we are only concerned with the effect that is created by the motion of the source. The Doppler effect is one of the basic characteristics of waves. Particles do not have this characteristic. For example, for a machine gun (source) that shoots 10 bullets per second, the speed or direction of motion of the gun does not affect the frequency of the bullets leaving the gun or received by a stationary target. The speed or the direction of the gun affects only the speed of each bullet.

It has been found that the Doppler effect is also one of the characteristics of light waves. For example the apparent light of stars that are advancing towards us has a higher frequency than that of retreating stars. Since light produces the Doppler effect, this suggests that light is a wave and not a particle.

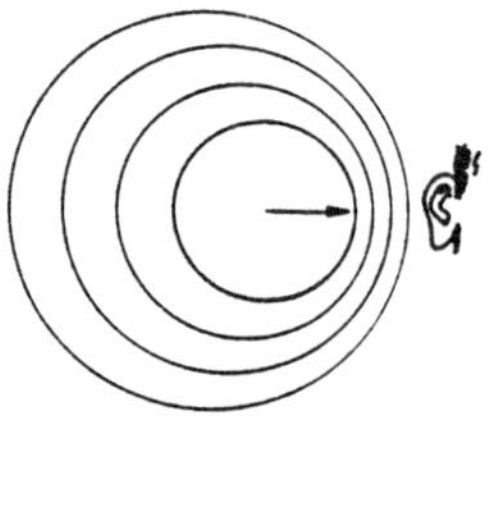

Fig. 1

APPENDIX F

Photon Theory: Evidence of Misrepresentation

Current light theories being taught in schools and universities are theories of contradiction. Scientists all agree that light theories are based on unexplained problems, but still continue to teach these theories because they came from Einstein. One such theory claims that light is a photon (corpuscle). Another, called the "electromagnetic theory" says that light is a wave. Einstein admits that these theories are opposite and contradictory. Here is what Einstein himself claimed:

> *One of the most fundamental questions raised by recent advances in science is how to reconcile the two contradictory views of matter and wave. It is one of those fundamental difficulties which, once formulated, must lead, in the long run, to scientific progress.*[247]

Scientists admit that these two theories are contradictory and impossible to reconcile into one theory. Nevertheless, they continue to teach both as if they were facts. Einstein was able to mislead the world by labelling these two contradictory theories "wave and matter duality theory of light". Before Einstein submitted his papers on relativity and the photon theory, all experiments, without exception, had indicated that light is a wave of ether. This beautiful, simple and confirmed fact explained everything without any difficulties until Einstein replaced it with two contradictory theories that not

---

[247] - Albert Einstein and Leopold Infeld, *The Evolution of Physics*, Simon and Schuster, New York, (1938), p.294.

only have not solved anything but have created many problems and contradictions.

Before we study the photon theory, it must be pointed out to the reader that the photon theory, the photoelectric effect formula and the new version of Maxwell's electromagnetic theory were all fabricated by Einstein and were all related to his relativity paper, in which he invented the idea that there is no need for ether. This is why he fabricated a postulate that light travels in empty space (without ether). Later, in support of his relativity ideas, he came up with the photon theory which eliminates ether altogether. In support of his photon theory, he presented a formula called the "photoelectric formula" which he claimed to be proof of the photon theory. One need not ask why, immediately following publication of his relativity paper, there suddenly appeared the photon theory, the photoelectric effect formula, and the new version of the electromagnetic theory, as evidence against ether. Soon after the publication of his relativity paper, Einstein was left with the problem of how to explain the nature of light without ether. His first attempt was to show that light is not a wave of any medium, but instead is made of corpuscles or photons that move in empty space. However, the photon theory failed to explain many wave-like phenomena suggesting that light is a wave. Hence Einstein misrepresented Maxwell's electromagnetic theory, which was developed on the basis of ether, to show that light is an electromagnetic wave that moves in empty space.

Before Einstein, all phenomena in physics indicated that light is a wave. After Einstein's fabricated misrepresentations, suddenly light became a particle. Careful examination of the photon theory reveals nothing to justify this theory's acceptance at all.

Centuries ago, there was a corpuscular theory for sound which led everyone to believe that sound was made of corpuscles; later sound was discovered to be a wave. Centuries ago, there was also a corpuscular theory for light which was proven to be wrong; light was found to be a wave. Here, Einstein has taken science backwards by making people believe light is a corpuscle. The following is "evidence" fabricated by Einstein in support of the photon theory and the relativity theory. It is called the photoelectric formula:

$$E = \tfrac{1}{2}mv + Q.$$

What is the Photoelectric Effect?

Here it will be illustrated that the photoelectric effect is not a proof that light is made of individual photons.

When a metal is heated, it may give off electrons. Some metals, when illuminated by light of sufficiently short wave- length, may also give off electrons. This phenomenon is called the "photoelectric effect".

In 1887, Heinrich Hertz noticed that a spark would jump more readily between two electrodes when their surfaces were illuminated by light. A year later, W. Hallwache investigated Hertz's findings and showed also that a clean, insulated zinc plate could acquire a positive charge when illuminated with ultraviolet light. He also showed that a negatively-charged plate when illuminated could lose its charge. He found the effect was enhanced in a vacuum. In 1900, Philipp Leonard proved that the effect observed by Hallwache was due to the separation of electrons from the metal surface as it is illuminated. In 1900, Elester and Geitel discovered that the number of electrons emitted per second is proportional to the intensity of light. In 1902, Leonard found that the maximum kinetic energy of electrons depends only on the frequency of light and is independent of its intensity. In 1905, Einstein, in support of the ideas in his relativity paper, claimed it was difficult to understand from Maxwell's wave theory how the kinetic energy of the electron could be independent of the intensity of light. Giving no explanation, he then claimed that the difficulty could be removed if we consider light to have a particle nature. In support of his idea, Einstein fabricated a formula called the "photoelectric effect formula" which he claimed supported the particle nature of light. Since a particle cannot have a wavelength, he also fabricated some assumptions that different photons must have different degrees of energy, and that a photon's energy must be proportional to the frequency of light. Again he gave no reason why this should be so.

After a very careful study of the photoelectric effect phenomenon and the formula that Einstein had presented as proof of the particle nature of light, it became very clear that the acceptance of the idea was based on superficial assumptions without any scientific proof. The following discussion will illustrate the reasons.

According to Einstein, light is made of corpuscles or photons that have a particle nature, and when a photon collides with an electron, the transfer of the photon's energy to the electron enables the electron to gain kinetic energy and to escape from the metal. In leaving the metal surface, the electron loses some of the energy that it had gained. According to Einstein:

$$E = \tfrac{1}{2}mv + Q$$

where E (the energy of a single photon) = $\tfrac{1}{2}mv$ (the kinetic energy of an electron) + Q (the energy used to liberate the electron). In other words, the formula says that part of the energy of a photon is converted into kinetic energy and part is used to liberate the electron. Einstein assumes that the formula proves there is a collision between a photon and an electron and that this in turn proves that light is made of individual particles (photons).

To understand how wrong and misleading Einstein's interpretations are, let us use an example. We know a car uses X amount of gas to accelerate from a speed of zero to a certain speed.
Then we can write:
$$E = \tfrac{1}{2}mv + Q$$

(Gas energy used) = (kinetic energy of the car) + Q
where Q is the sum of energy wasted in the exhaust plus the energy dissipated as friction, heat, resistance to air, etc. To simplify, we can say that part of the gas energy is converted into kinetic energy and part as waste.

Let us assume that .0051 litre of gas is used to accelerate a 1000 kilogram car to 100 kilometers per hour. The photoelectric formula assumes the energy gained by an electron is the unit of energy. In other words, the example above assumes that the amount of gas used (.0051 L) is the unit of gas energy. The fact is this: .0051 litre of gas is a combination of many drops.

In physics, there is a phenomenon called "resonance". For resonance to occur, only the frequency counts, not the intensity. For example, for a musical instrument such as a violin, its strings can be excited by resonance when the sound of another instrument with exactly the same pitch (frequency) reaches the string. In this case, the material body, being the string, is excited by waves which have the same frequency as that of the natural frequency of the string. Similarly, for electrons to be excited by resonance, the light waves must have the same frequency as the orbital motion of the samareh.

Einstein assumes that light is made of photons and only one photon is responsible for the entire kinetic energy of the electron. Such assumptions are superficial and are based on a formula that is so general that it can never be called a proof or an evidence for his particular case at all.

Einstein and relativity supporters have used a phenomenon as an evidence in support of the photon idea. During an eclipse of the sun, the light of stars that appear very close to the sun bends in the direction of the sun. Einstein and relativity supporters have reasoned that the light of the stars passing by the sun is bent because the gravity of the sun attracts the photons.

Here is another explanation: this phenomenon indicates that the sun has its own atmosphere. As the light of nearby stars moves through the sun's atmosphere their light waves are diffracted by the change in the density of the medium.[248]

The following are some of the assumptions made in the photoelectric formula by Einstein, together with a discussion of their weaknesses:

1 - The photoelectric formula assumes that electrons, before release from the atom, have a speed of zero. They are assumed to be in a stationary state waiting to move only after a collision with a photon. The fact is, however, that only in absolute zero temperature is an electron at rest, and that the higher the temperature, the faster the motion of electrons

---

[248] - See Otto Luther, *Relativity Is Dead*, California, Key Research Corp., (1986), which explains the reason in detail and gives an example of a fish bowl that diffracts the light ray passing through the side of the bowl.

around the nucleus and the less energy they need to escape from the metal surface. According to the photoelectric formula, an electron, having no initial velocity, gets all of its velocity from a collision with one photon.

2 - It assumes the light energy to be in an individual packet called a photon.

3 - It assumes the packet (photon) to be of particles.

4 - It assumes the motion of only one hypothetical packet of matter to be responsible for the entire kinetic energy of an electron. It overlooks the fact that in each second over 10^{14} wave fronts, one after the other, reach the electron. To understand how large this number is and how short the time between two consecutive wave fronts, if we stretched out the time between the two consecutive wave fronts into one second of time, then proportionally it is as if we have stretched out one second of time into 3.1 million years. This is because in 3.1 million years there are about 10^{14} seconds, and in one second about 10^{14} wave fronts reach the electron. Einstein, without any evidence, assumes that only one photon is responsible for the entire kinetic energy of the electron.

It must be borne in mind that every electronic instrument, such as a Geiger counter, functions by the number of electrons and not the individual photons or wave fronts. So far, there is no instrument invented that can count precisely the individual number of wave fronts or photons that an electron receives. Where is the evidence to show that only one photon could be responsible for the entire kinetic energy?

5 - It assumes that the amount of energy gained by the electron is a minimum unit (photon) of light energy. Einstein ignores the fact that the amount of energy an absorbing surface receives depends on how large the absorbing surface is. For example, a large surface exposed to the light of the sun receives more light energy than a small surface. If the size of the electron were much larger, it would gain much more energy.

6 - It assumes that the hypothetical photon, after a collision, becomes energy; and that the photon, which is supposed to be a particle, suddenly and for no reason, disappears and becomes non-existent.

7 - It assumes a general formula that can represent the energy conversion of innumerably different phenomena to have only one single interpretation.

8 - It assumes that the interpretation is a sufficient proof that light is a particle, on the hypothetical basis that all of the above assumptions are correct and that the following assumptions are also correct.

The following are a few reasons proving the impossibility that light is composed of individual photons. On the contrary, there is no light phenomenon that the wave nature has not explained with simplicity and satisfaction.

1 - Einstein, giving no reason, advances a claim that the photoelectric effect cannot be explained by the wave nature of light. The truth is that there are many examples which show that wave energy does indeed convert to kinetic energy. When a jet breaks the sound barrier, it is the sound wave that breaks the windows and sends pieces of glass

flying. When a bomb explodes, it is the shock wave that causes the destruction and sends the debris flying.

Altberg has shown that a light wave does create a pressure on reflecting surfaces. He says:

> *The existence of pressure on a surface due to the incidence of a normal beam of light, first deduced a consequence of the electromagnetic theory by Maxwell, has been fully confirmed by the experiments of Lebedew by Nicholas and Hull. These experiments show that the pressure exists and that it is equal to the energy per c.c. or to the energy density in the incident beam.* [249]

It must be noted that this deduction was based on Maxwell's wave theory and not on the hypothetical photon theory. Maxwell showed that the transmission of light waves in the medium of ether which exists in space and the transmission of electromagnetic disturbances in the medium which exist in a conductor are both in essence the same phenomenon. In other words, both light and electric signals are transmitted by the same medium of ether.

According to Maxwell's electromagnetic theory, the large spaces between the atoms and electrons of a conducting wire are not empty but are filled with the ethereal medium,[250] and the wave of this medium creates the current of electrons in the conductor. The speed of the wave of the medium (electric signal) along the wire was found to be the same as the speed of light, but the speeds of the motion of

---

[249] - *Philosophical Magazine*, Ser. 6.9, (1905), p. 169.

[250] - See Maxwell, "A Dynamical Theory of the Electromagnetic Field", fn. 50, p. 460, (4).

electrons along the conductor were only a small fraction of the speed of light.

There is a parallel similarity between the current generated by light in the photoelectric effect and the current generated along a conductor by an electromagnetic disturbance. Both light waves and electromagnetic disturbances along a conductor generate electrical current. Maxwell proposed that the reason for the similarity is because both are waves of the same medium, differing only in frequency or wavelength.

2 - If light is made of photons, then there is no explanation for why the photon emissions in all directions, side by side, are so perfectly organised and create a perfect wave-plane front.

3 - If we assume that light is made of photons, then there is no explanation for the phenomenon of interference. No such explanation has been found despite the intense effort by Einstein and his group to find some even superficial explanation. On the other hand, the wave nature of light, with total simplicity, explains it all, because interference is one of the fundamental characteristics of waves.

In 1902 O. Lummer and E. Gehrcke obtained interference 1 meter apart from the light of a star. In 1920 at the Mount Wilson Observatory, these same two scientists proved that it is impossible that the interference which they obtained from the very large coherent beams of a star 20 meters apart could be a single corpuscle or photon.[251]

---

[251] - See Sir Edmund Whittaker, *A History of the Theories of Aether and Electricity*, New York, (1962), p. 95.

4 - If we assume that light is made of photons, no explanation can be found for the phenomenon of the reflection of light from a surface. For example, if light were made of photons, we should not be able to see our image in the mirror at all, because the size of photons would be so small that they would pass through the atoms of the mirror's surface. Any shadowy reflection in the mirror would depend on haphazard collisions of individual photons with individual subatomic particles. The small percentage of reflection would be in all directions depending on how and on which side of the particle each collision occurred.

5 - If we assume that light is made of photons, there is no explanation for why the different colours of light all travel with the same speed. On the contrary, light being a wave of ether explains everything beautifully. The fact is that all waves, such as water, sound or light waves, are transmitted by their medium. The speed of the wave depends on the nature of the medium which carries the wave and not the frequency.

6 - If we assume that light is made of photons, then why is the speed of such photons independent of the velocity of their source, especially if space is considered empty, without ether? It is a known fact that the velocity of any wave is always independent of the velocity of its source and that it is only dependent on the nature and condition of the medium that carries the wave. If the photon is a particle, according to the laws of physics, the velocity of the photon should depend on the velocity of the source.

7 - If we assume that light is made of photons with a particle nature, why does the photon not slow down when it collides with a surface, such as the surface of a mirror? However,

light as waves answers everything, since one of the basic characteristics of any wave is that it travels with a constant velocity regardless of whether it is reflected from a surface or not. This is the case with any wave such as sound or water, because the waves are carried by the medium, and the velocity of the wave depends only on the qualities of the medium and not on the source or reflective surface that acts as a secondary source.

8 - If we assume that light is made of photons, why do photons that move in opposite directions never collide? If a photon can have collisions with electrons, why not have collisions with another photon? If we assume that light is made of photons, then why do lights that move in opposite directions behave only as waves and not particles?

9 - If we assume that light is a particle, then why does light, after passing through water or glass into open air, have a higher speed than the speed it had in water or glass? The fact is this: if light was a particle, it would not have increased its speed in air because for a particle to increase its speed it must receive energy from somewhere. Where and how do photons receive energy? The phenomenon can only be explained by the wave nature of light because light is carried by a medium, and the speed of the wave only depends on the quality of the medium. Light that moves slowly in glass after passing through the glass is carried by a medium which has less density. This phenomenon in itself indicates that light is a wave and not a particle.

10 - When an electron jumps between two electrodes, it generates a spark. If we assume the spark is a series of photons leaving the electron, then the electron during its travel from one wire to another must give off zillions of

photons in all directions. Even if all of the electrons were converted to photons, a small fraction of that many photons could not be produced. Thus the question arises, "Where do these photons come from?" On the other hand, the understanding that an electron's motion in ether generates friction and waves that move in all directions poses no problems.

11 - Why can no related phenomena be found that do not have contradictions with the photon theory?

12 - Why do all related phenomena agree with the wave nature of light?

Here we see that the photoelectric effect formula which allegedly "proves" Einstein's photon theory is without scientific foundation. All these reasons illustrate how superficial has been Einstein's so-called evidence against ether.

APPENDIX G

Einstein's Misrepresentation of Maxwell's Theory: Evidence of Fraud

In previous chapters, we studied some of the problems with the photon theory. Here we are going to examine the problems of the electromagnetic theory.

Before we study the theory itself, it must be pointed out that there is a controversy among physicists about the basic assumptions in Maxwell's electromagnetic theory. On the one hand, some who have read Maxwell's papers believe that his theory is based entirely on the medium of ether and that, according to the theory, light is an undulation (wave or vibration) of ether. On the other hand, the version shown in textbooks and scientific publications, based on Einstein's misrepresentation of Maxwell's theory,[252] states that light is an electromagnetic wave that spreads in empty space.

Here we are going to study Maxwell's actual paper together with Einstein's version. Then it will be clear to the reader that what we read in Einstein's papers or books has nothing to do with Maxwell's actual theory at all.

---

[252] - Albert Einstein and Leopold Infeld, *The Evolution of Physics*, Simon and Schuster, New York, (1938), p. 155.

Maxwell's Electromagnetic Theory According to Einstein

The following is how Maxwell's theory has been misrepresented by Einstein.[253]

<u>What is a magnetic field?</u>
Here is what Einstein writes:

> *If, for instance, a magnet attracts a piece of iron, we cannot be content to regard this as meaning that the magnet acts directly on the iron through the intermediate empty space, but we are constrained to imagine - after the manner of Faraday - that the magnet always calls into being something physically real in the space around it, that something being what we call a "magnetic field.' In its turn this magnetic field operates on the piece of iron, so that the latter strives to move towards the magnet. We shall not discuss here the justification for this incidental conception, which is indeed a somewhat arbitrary one.*[254]

Einstein claims that the space around a magnet is a magnetic field. A piece of iron in a magnetic field will experience a force. How the force can act without any material connection is not known. No one has been able to

---

[253] - See Einstein's 1905 Relativity Paper; see also Einstein and Infeld, *The Evolution of Physics,* fn. 18, p. 129-156. See also *Does the Inertia of a Body Depend on its Energy Content?*

[254] - Albert Einstein, *Relativity, the Special & the General Theory,* Translated by R. W. Lawson, 3rd ed, Methuen & Co. Ltd. London, (1920), p. 63.

explain what kind of force or energy that comes out of a magnet can attract or repel an object.

We know that a suction hose that sucks air creates a force of attraction. We know also that an air pressure hose from which air is coming out creates a repulsive force. In these cases, the current of air is responsible for the forces of attraction and repulsion. The flow of air is the material connection responsible for creating a force that acts at a distance.

In the case of a magnet, if we assume that there is no current of ether, then how is a force created that can, in one case, attract a piece of iron , and in another case, repulse another magnet? Einstein does not provide an answer as to how this can be possible. Compare with the ether explanation for magnetism in Chapter 7.

What is an electric field?

The space around a charged body is an electric field where another charged body experiences a force. According to Einstein, this force is of action-at-a-distance type, making itself felt without the presence of any material connection whatsoever.[255] Einstein does not provide any answer as to how this can be possible.

What is an electromagnetic wave?

According to Einstein's interpretation of Maxwell's theory, if either an electric or magnetic field is moving in space or changing in intensity, the motion or change sets up the other field, that is, a changing electric field sets up a magnetic field, and vice versa. This interrelationship between magnetic and electric fields creates electromagnetic waves by

---

[255] - See Einstein & Infield, *The Evolution of Physics*, fn. 18, p.152.

which light travels or radio communication is carried out, for such waves are simply travelling fields in which the energy is alternately handed back and forth between the electric and magnetic fields. Here is what Einstein wrote:

> *What kind of changes are now spreading in the case of an electromagnetic wave? Just the changes of an electromagnetic field! Every change of an electric field produces a magnetic field; every change of this magnetic field produces an electric field; every change..., and so on. As field represents energy, all these changes spreading out in space, with a definite velocity, produce a wave.* [256]
> *The electromagnetic waves spreads in empty space.*[257]

THE FOLLOWING ARE SOME OF THE ASSUMPTIONS THAT ARE FOUND IN EINSTEIN'S CLAIMS

1. The electric field is that part of space in which the charge can act at a distance without any material connection.
2. The magnetic field is that part of space near a magnet in which another magnet experiences a force without any material connection.
3. The force needs no material connection to act at a distance. Hence there is no motion of matter to cause action at a distance.
4. The fields can change from magnetic to electric with no explanation as to how and why they can do so.
5. Because the field changes from one kind into another, it can travel from one location into a new location. In other words, the change in nature is the means of locomotion.

---

[256] - Ibid, p.154.

[257] - Ibid, p.151.

Einstein does not provide any answer as to why the electric field must change its nature in order to travel in space.
6. Light is simply a travelling field.
7. No matter how fast or how slow the field changes from electric to magnetic and vice versa, the field travels in empty space with a definite velocity.
8. Light is an electromagnetic wave.
9. Light travels without the medium of ether. There is no need for the medium of ether to transmit electromagnetic waves.

WHAT ARE THE PROBLEMS WITH THESE ASSUMPTIONS?

The following are some of the main problems with these assumptions. With a very careful study of Maxwell's actual paper, I found that none of the above assumptions exist in Maxwell's theory. In fact, I found that all are contrary to his theory. Here are some examples.

According to Einstein, one of Maxwell's assumptions is that the force or field acts without any medium or material connection, whereas in Maxwell's paper, we find that the force or field is produced by:

> . . . *an action which goes on in the surrounding medium as well as in the existed bodies and endeavouring to explain the act between distant bodies without assuming the existence of forces capable of acting directly at sensible distances.* [258]

---

[258] - See Maxwell, "A Dynamical Theory of the Electromagnetic Field", fn. 50, p. 460, (2).

According to Einstein, the electric or magnetic forces are capable of acting directly at sensible distances, whereas in Maxwell's theory such assumptions are set aside altogether: "*without assuming the existence of forces capable of acting directly at sensible distances.*"[259]

An electric or magnetic field, according to Einstein, is that part of space in which electric or magnetic forces can act (without the need for a material connection), but, in Maxwell's actual paper, an electric or magnetic field is defined as that part of space where matter is in motion and by which force is produced. [260]

According to Einstein, the undulation of light takes place in empty space without ether, whereas in Maxwell's theory, space is always filled with matter. According to Maxwell, if we remove from a space all the gross matter such as air or gas molecules, in such a vacuum there remains enough matter to receive and transmit the undulations of light.

> *The electromagnetic field is that part of space which contains and surrounds bodies in electric or magnetic conditions.*
>
> *It may be filled with any kind of matter, or we may endeavour to render it empty of all gross matter, as in the case of Geissler's tubes and other so-called vacua.*
>
> *There is always, however, enough of matter left to receive and transmit the undulations of light and heat, and it is because the transmission of these ra-*

---

[259] - Ibid. p. 460, (2).

[260] - Ibid. p. 460, (3).

> *diations is not greatly altered when transparent bodies of measurable density are substituted for the so-called vacuum, that we are obliged to admit that the undulations are those of an œthereal substance, and not of the gross matter, the presence of which merely modifies in some way the motion of the œther. We have therefore some reason to believe, from the phenomena of light and heat, that there is an œthereal medium filling space and permeating bodies, capable of being set in motion and of transmitting that motion from one part to another, and of communicating that motion to gross matter so as to heat it and affect it in various ways.*[261]

It must be born in mind that *œther* is an old way of saying ether.

According to Einstein, light is a travelling wave that can move in empty space, whereas, according to Maxwell's actual theory, light is electromagnetic disturbances of the medium of ether, disturbances which travel through the field (the "*...action which goes on in the surrounding medium*").

> *The agreement of the result seems to show that light and magnetism are affection of the same substance propagated through the field according to electromagnetic laws.*[262]

In Einstein's papers there is no mention of matter and how the electric or magnetic forces are created, whereas in

---

[261] - Ibid. p. 460, (4).

[262] - Ibid. p. 499, (97).

Maxwell's theory these forces are created by the motion of etherial matter:

> *The theory I propose may there be called a theory of the "Electromagnetic field", because it has to do with the space in the neighbourhood of the electric or magnetic bodies, and it may be called a "Dynamical" Theory, because it assumes that in that space there is matter in motion, by which the observed electromagnetic phenomena are produced.*[263]
>
> *According to the theory which I propose to explain, this "electromotive force" is the force called into play during the communication of motion from one part of the medium to another.*[264]

According to Einstein, the propagation of light wave "*undulation*" *consists of a continual change of energy from one form into another, that is to say, from magnetic energy into electric and from electric into magnetic energy*, without any explanation of why such changes occur. However, in Maxwell's actual paper, the propagation of light undulation (wave) occurs because of the medium:[265]

> *The medium is capable of the receiving and storing of two kinds of energy, namely the "actual" energy depending on the motion of its parts and "potential" energy...*

---

[263] - Ibid. p. 460, (2).

[264] - Ibid. p. 461, (10).

[265] - Ibid. p. 461, (7).

According to Einstein, since a changing magnetic field generates a changing electric field and vice versa, the fields are able to travel in space. There is no explanation as to how this is possible; nor is there any explanation of how the change is responsible for the field's travelling in space. What has the change in form of energy to do with the change in its location? There is no phenomenon in nature that can be used as a parallel for such a strange and unreasonable hypothesis. According to Maxwell's actual theory, the medium of ether is responsible for the transmission of wave energy. As the medium of air is responsible for the transmission of sound energy, the change of energy from kinetic into potential and vice versa can be easily transmitted by a medium.

According to Einstein, light is an electromagnetic wave, whereas in Maxwell's paper, light is described as an electromagnetic disturbance.[266]

The following is a summary of the problems with Einstein's presentations of Maxwell's theory which are reflected in current physics textbooks:
1. All the assumptions in Einstein's papers are contrary to Maxwell's assumptions in his actual paper.
2. Einstein's interpretation does not explain any fundamental questions.
3. Einstein's interpretation does not explain how force acts at a distance.
4. Einstein's interpretation does not explain how a field travels in space.
5. Einstein's interpretation does not explain how energy can change its form; why a disturbance created by a changing

---

[266] - Ibid. p. 499, (97).

magnetic field sets up a changing electric field; or why a disturbance created by a changing electric field sets up a changing magnetic field.
6. Einstein's interpretation does not explain why the velocity of the travelling field in empty space is always the same.
7. Einstein does not explain why the velocity of the wave is independent of how fast the fields interchange from one kind into another.
8. Maxwell's wording of electromagnetic disturbance has been misinterpreted to mean "electromagnetic wave".[267]
9. In Einstein's version, the essential part of Maxwell's theory is missing.

A very quick look at Maxwell's original work shows that he (page 2) has developed a groundwork for his theory, including: definitions of ether, light and the magnetic field. On this ground, Maxwell worked out his theory.

Here is an analogy to show why Maxwell needed such groundwork. Everyone has seen a contract agreement for the purchase and sale of a house. In the beginning of the agreement, there is always groundwork, to identify the purpose of the contract, the purchaser, the vendor, and what the specifications and address of the house are. After the groundwork, the rest of the contract does not repeat information contained in the groundwork. For example, in the rest of the contract, there is no specification or address of the house, but instead there is the word "premises". This is done to shorten the contractual agreement and to avoid repetition of information. Similarly, in Maxwell's paper, one sees no repetition of the information contained in the groundwork. For example, in the groundwork, Maxwell

---

[267] - Albert Einstein, *Relativity, the Special & the General Theory*, Translated by R. W. Lawson, 3rd ed, Methuen & Co. Ltd. London, (1920), p. 63.

clearly specifies the nature of the magnetic field and how and why the force acts as it does. In the rest of the paper, there is no repetition of this information. Instead, Maxwell uses the term "Magnetic field". Imagine if in Maxwell's paper, with every use of the word "field", the description of the nature of the fields that are found in the groundwork was also repeated. How long and unprofessional the paper would become!

What happens if the groundwork is missing from the contract agreement? The whole contract then becomes useless because the essentials are missing: no one knows who is selling to whom and what is being sold. Similarly, without the groundwork, Maxwell's paper is an incomplete and dead theory because it lacks essential information.

For example, no one knows what a magnetic field is, why in a magnetic field there is a force, why the force makes itself felt at a distance, what the electromagnetic field means, why the electromagnetic disturbances propagate as waves in space, why the velocity of the propagation is always the same, or why the velocity of the wave is independent of the frequency of the wave. All these essential bits of information are missing.

Einstein presented Maxwell's theory without its groundwork. Many textbooks and almost all scientific publications have done the same. Einstein presented Maxwell's theory without the essential ingredient of ether which is in the groundwork. If Maxwell meant what Einstein and his supporters meant, why is there no indication of it in all of Maxwell's papers? If Maxwell had concluded that light is only an electromagnetic wave without the medium of ether, then why in all of his papers did Maxwell never use the words "electromagnetic wave"? Instead he always referred to an electromagnetic disturbance, and to the medium; that light and magnetism are "affections" of one and the same medium, or that light is the undulation of ethereal matter in a so-called vacuum; that light is an electromagnetic disturbance of the medium.

Why can none of the basic assumptions which we find in Einstein's papers, books and in the scientific publications of today be found anywhere in Maxwell's works? From where did he derive his assumptions?

In an article in the Encyclopaedia Britannica printed in 1890, Maxwell himself has written an article under the section "Ether" which included his electromagnetic theory of light. After describing the function of the "...*æther in electromagnetic phenomenon...*" he describes his Electromagnetic Theory of Light:

> *Function of the æther in electromagnetic phenomena..*
> *- Faraday conjectured that the same medium which is concerned in the propagation of light might also be the agent in electromagnetic phenomena.* "*For my own part,*" *he says,* "*considering the relation of a vacuum to the magnetic force, and the general character of magnetic phenomena external to the magnet, I am much more inclined to the notion that in the transmission of the force there is such an action, external to the magnet, than that the effects are merely attraction and repulsion at a distance. Such an action may be a function of the æther; for it is not unlikely that, if there be an æther, it should have other uses than simply the conveyance of radiation.*"[4]
> *This conjecture has only been strengthened by subsequent investigations.*

....

Electromagnetic Theory of Light. - *The properties of the electromagnetic medium are therefore as far as we have gone similar to those of the luminiferous medium, but the best way to compare them is to determine the velocity with which an electromagnetic disturbance would be propagated through the medium. If this should be equal to the velocity of light, we would have strong reason to believe that the two media, occupying as they do the same space, are really identical.*

....

Physical constitution of the œther. - *What is the ultimate constitution of the œther? Is it molecular or continuous?*

We know that the œther transmits transverse vibration to very great distances without sensible loss of energy by dissipation.[268]

[4] - Experimental researches, 3075.

Therefore it is very clear that Einstein misrepresented Maxwell's theory.

---

[268] - *Encyclopedia Britannica*, ninth edition, Vol. VIII, (1890), p. 571.

APPENDIX H

"THE RELATIVITY OF SIMULTANEITY[269]"

Up to now our considerations have been referred to a particular body of reference, which we have styled a "railway embankment." We suppose a very long train travelling along the rails with the constant velocity v and in the direction indicated in Fig. 1. People travelling in this train.........

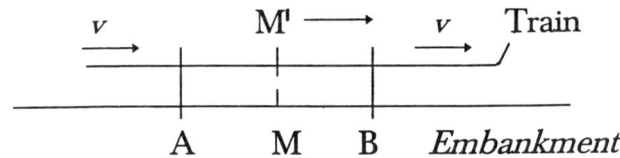

Fig. 1

Are two events (e.g. the two strokes of lightening A and B) which are simultaneous with reference to the railway embankment also simultaneous relatively to the train? We shall show directly that the answer must be negative.

When we say that the lightning strokes A and B are simultaneous with respect to the embankment,

---

[269] - Albert Einstein, *Relativity, the Special & the General Theory*, Translated by R. W. Lawson, 3rd ed, Methuen & Co. Ltd. London, (1920), pages 25-26.

we mean: the rays of light emitted at the places A and B, where the lightning occurs, meet each other at the mid-point M of the length $A \rightarrow B$ of the embankment. But the events A and B also correspond to positions A and B on the train. Let M' be the mid-point of the distance $A \rightarrow B$ on the travelling train. Just when the flashes of lightning occur, this point M' naturally coincides with the point M, but it moves towards the right in the diagram with the velocity v of the train. If an observer sitting in the position M' in the train did not possess this velocity, then he would remain permanently at M, and the light rays emitted by the flashes of lightning A and B would reach him simultaneously, i. e. they would meet just where he is situated. Now in reality (considered with reference to the railway embankment) he is hastening towards the beam of light coming from A. Hence the observer will see the beam of light emitted from B earlier than he will see that emitted from A. Observer who takes the railway train as their reference-body must therefore come to the conclusion that the lightning flash B took place earlier than the lightning flash A. We thus arrive at the important result:

Events which are simultaneous with reference to the embankment are not simultaneous with respect to the train, and vice versa......

1- As judged from the embankment.

APPENDIX I

Longitudinal and Transverse Waves of Light

Today the general belief is that all light waves are transverse waves and no light is longitudinal. The belief is mainly based on the phenomenon called "polarisation". In order to see why scientists arrived at the idea that light is a transverse wave, let us examine the phenomenon of polarisation and the history of its discoveries.

In 1669, E. Bartholin observed that objects viewed through a certain type of crystal appeared double. The phenomenon was called "double refraction." because an incident ray is split into two rays by two different refractive qualities of the crystal. Later, Christian Huygens repeated the experiment and directed the rays emerging from one crystal towards a second crystal. He found that the second crystal must be placed in a certain position in order for the rays to pass through it. In 1908, E.L. Malus found that light reflected under certain angles of incidence exhibited the same properties as rays that had passed through any of the above crystals. Malus for the first time used the term "polarisation" because he believed that light is made of long corpuscles, each having two poles, similar to a magnet. When a beam of corpuscles passes through the crystals, they are aligned or polarised in a certain plane. In 1816, F. Arago and A. J. Fresnel discovered that two light beams whose polarisation planes are perpendicular to each other do not interfere. However, two light beams whose polarisation planes are parallel to each other do produce interference fringes. Based on these findings, T. Young proposed that light waves are transverse waves in ether. Faraday, who was a strong supporter of the ether theory, proved that his magnets could turn the plane of polarisation of a light beam through a certain angle. He also proved that the direction of rotation of the beam of light depends upon the polarity of his magnet.

First of all, a particular experiment in which light becomes polarised cannot be the basis to claim that all light is composed of transverse waves and not longitudinal. Physicists recognise that when light passes through a certain medium such as a crystal, it is forced to behave in a particular way. To cite the understanding of present day scientists:

If a beam of light passes through certain crystals, the systematic arrangement of atoms in the crystal forces the light beam to oscillate in a particular plane. This oscillation in a single plane is called Polarisation.[270]

Secondly, at the time Malus made the discovery of polarisation, scientists had no idea about the atomic structure of crystals. At the time, the reason for polarisation was attributed to light itself rather than to the reflecting surface of the crystal which he used. Although years later Sir David Brewster made investigations over a wide range and realised that the angle of polarisation has something to do with the refractive index of the reflecting medium (crystal), even this discovery did not direct the attention of scientists to the atomic structure of the medium. However, today the knowledge of the qualities and characteristics of the mediums or crystals have immensely increased to the point where, if one investigates phenomena such as polarisation, lasers and how crystals can change the quality and character of a light beam, one can clearly see that polarisation also has

---

[270] - *The Realm of Science*, Touchstone Publishing Company, Louisville Kentucky, (1972), v. 6, p.103.

something to do with sorting out the phase difference in the incident light. Furthermore, some crystals function only within a certain range of frequencies. For example, quartz crystals are used to sort out the frequencies in radio transmitters and receivers.

Furthermore, the fact that the plane of crystal must be turned, for example, 90 degrees, to show or reflect a light beam, indicates that there is a possibility that the phenomenon has something to do with the internal structure of the crystal and how the unit faces of each crystalline are arranged side by side within it. Each atom within the crystal could act like a mirror reflecting at 90 degrees. When the crystal is arranged at a certain angle, the reflection from one atom to another enables us to see the light.

It is interesting to note that since the early nineteenth century it has been claimed that there can be no transverse waves in air or water. This claim is questionable, for it is difficult to verify whether a sound wave is transverse or longitudinal; there are no instruments that I know of such as polarisers and analysers which can distinguish between transverse and longitudinal sound waves.

APPENDIX J

Mechanical View of Ether

Based on the idea that light is a transverse wave, nineteenth century physicists concluded that ether must be like jelly. This mechanical view of ether as jelly not only had no negative effects on their belief about the existence of ether, but actually enhanced it. Here is what Hertz, who discovered the radio wave, said:

> *What then is light? Since the time of Young and Fresnel we know that it is a wave-motion. We know the velocity of the waves, we know their wavelength, we know that they are transversal waves; in short, we know completely the geometrical relations of the motion. To the physicist it is inconceivable that this view should be refuted; we can no longer entertain any doubt about the matter. It is morally certain that the wave theory of light is true, and the conclusions that necessarily follow from it are equally certain. It is therefore certain that all space known to us is not empty, but is filled with a substance, the ether, which can be thrown into vibration ...*[271]

If one investigates some of the jelly-like oil products on the market, one notices that some of them have so little viscosity that if one's hand is immersed in the jelly, one can hardly feel its friction. However, if a bowl full of the jelly is shaken slightly, one can clearly see how a transverse disturbance can create waves which move in the jelly.

---

[271] - Miscellaneous papers by Henrich Hertz with an introduction by Professor Phillop Lenard, 1896, p. 313, 314.

Comparing the quality of this kind of jelly with that of ether, one realises that, first of all, the ether is millions of times less dense than the jelly. Secondly, ether is also millions of times less rigid. This explains why ether can transmit transverse waves. This also explains how the planets move through ether without noticeable resistance. It must be born in mind that, regarding all the planets orbiting the sun, the spin of the sun and the spiral motion of the solar wind (ether) counterbalances the resistance of ether.

Before Einstein, scientists had calculated the coefficient of rigidity of ether and found that the calculated amount of elasticity was reasonable, with no problems. However, Einstein changed everything. In order to prove that there is no ether, he claimed that ether as jelly does not fit the mechanical view. Since he had no scientific evidence for his claim, he wrote:

> *In order to construct the ether as a jelly-like mechanical substance physicists had to make some highly artificial and unnatural assumptions. We shall not quote them here, they belong to the almost forgotten past. But the result was significant and important. The artificial character of these assumptions, the necessity for introducing so many of them quite independent of each other, was enough to shatter the belief in the mechanical point of view.*[272]

Today one of the arguments that fourth dimension supporters use to disprove ether is that if ether exists, it should be extremely rigid. In order for light to travel through a medium with a high velocity of 300,000 kilometers per

---

[272] - See Einstein & Infield, *The Evolution of Physics,* fn. 18, p.123.

second the medium should be extremely rigid. They reasoned that the more rigid a substance is, the higher the velocity of sound travelling through it. For example, the velocity of sound in steel is much higher than in air. They claimed that in order for light to travel in ether with a velocity of 300,000 kilometers per second, the ether must be many thousands of times more rigid than steel. This kind of reasoning used against ether is like saying that since the ocean waves move with a velocity of 3 kilometers per hour, imagine how much more rigid the air must be in order for sound waves to move with the velocity of 1200 kilometers per hour.

The supporters of the fourth dimension ideas have ignored the fact that the rigidity of substances such as steel depends on the molecular bindings which, in turn, depends on the motions of the electrons and their orbits. The ether substance transmitting light waves does not have molecular binding, does not have electrons orbiting the ether particles. Furthermore, they have also ignored the fact that in the same substance such as water or steel, different kinds of waves can travel with different velocities. For example, the velocity of the water waves over the oceans is only a few kilometres per hour whereas the velocity of sound in the same water is several hundred times greater. The wavelength of the ocean waves is in the order of 100 meters, whereas the wavelength of sound waves in the same water is about 1,000,000 times smaller. Light which has a wavelength of 100,000,000,000 times smaller than sound waves moves with a velocity of about 300,000,000 meters per second in ether, which permeates the water.

Index

—A—

abandonment of ether, 161, 255
aberration, 13, 15, 19, 137, 141, 142, 143, 148, 149, 150, 151, 202, 214, 221, 255, 263
abuse of mathematics, 67, 75
Airy, 148
Alfven, 129
alpha particles, 50, 51, 254, 307
Altberg, 274
alternating current, 84
Ampere, 41
angular momentum, 78
art, 236, 238, 255
asteroid, 114, 133
asteroids, 132
atomic clock experiment, 222, 234, 235
Aurora, 115
avalanche, 45

—B—

Beckmann, 217, 228
Bell, 132
beta particles, 49
Bohr, 174, 177
Bothezat, 243
bow shock, 20, 93, 94, 96, 97, 101, 102, 111, 112, 137, 139, 140, 153, 154, 203, 226, 255
Bradley, 13, 142, 145, 147, 263
Brahic, 103
British committee, 177
BrownIan, 228
Bush, 176

—C—

chain reaction, 163, 173, 174, 175
Christian Huygens, 13, 294

Churchill, 177
coma, 92, 96, 97, 99, 101, 103, 116, 136, 137, 138, 255
comet, 8
comets, 9, 46, 83, 92, 96, 97, 98, 99, 100, 101, 102, 111, 113, 114, 115, 135, 137, 138, 252, 254
Comptom, 176
contraction, 16, 17, 18, 184, 186, 187, 189, 191, 215, 216, 221, 222, 226
corona discharge, 30
corpuscular theory, 13, 14, 18, 171, 262, 267
Courtivron, 13
Curie, 22, 173
curvature of time and space, 70, 71

—D—

Davison, 46
density, 13, 23, 33, 34, 117, 119, 121, 130, 132, 164, 191, 231, 261, 271, 274, 277, 284
Dingle, 73, 74, 75, 76, 169, 228, 233
Doppler effect, 29, 222, 252, 264, 265
drag of ether, 19, 145, 147, 148, 152, 215, 255
drift velocity', 90

—E—

earth, 13, 15, 16, 19, 34, 35, 36, 44, 46, 48, 53, 59, 60, 61, 62, 63, 64 - 68, 70, 78, 79, 94, 99, 102, 110 - 115, 119, 121, 122, 125, 126, 127, 128, 129, 130, 132, 135, 137 - 143, 145 - 148, 150 - 156, 184, 192, 193, 198, 199, 201, 202, 203, 214, 216, 221, 226, 227, 232, 234, 241, 243, 245, 254, 263, 264, 305, 306
earth drags ether, 20, 139, 140, 142, 147, 148, 153, 202, 227
eclipse, 13, 125, 130, 271
Einstein, 17, 18, 19, 20, 21, 25, 31, 32, 44, 46, 67 - 76, 81, 82, 87, 132, 135, 140, 142, 149, 150, 152, 154, 156, 158 - 178, 180- 196, 199- 209, 211-236, 239 - 245, 251, 254 - 256, 258, 262, 264, 266 - 273, 275, 278 - 289, 291, 292, 298
electric field, 32, 80, 81, 88, 197, 254, 281, 282, 283, 286, 287
electric fields, 82, 84, 282
electricity, 21, 38, 76, 79, 89, 160, 200, 207, 220, 242, 245, 248, 249, 251
electromagnetic theory, 19, 20, 89, 205, 207, 208, 209, 210, 225, 256, 266, 274, 279, 289
electromagnetic waves, 82, 208, 211, 275, 282, 283
electromoon, 36, 44
electron, 8, 30, 36, 37, 38, 39, 43, 44 - 47, 49, 78, 79, 80, 81, 82, 84, 87 - 92, 181, 185, 223, 225, 251, 252, 254, 268 - 273, 277
electron diffraction, 8, 38
electron wave, 8, 38, 48

320

Elester, 268
eruption, 124, 126
Essen, 217, 228
ether propulsion, 249, 250
Ether Propulsion System, 245
ether spacecraft, 250
ether wind, 16, 20, 34, 126, 140, 151, 152, 154, 155, 156, 203, 226
Euler, 13

—F—

Faraday, 14, 31, 212, 213, 251, 280, 290, 294
fission products, 25
FitzGerald, 16, 184
Fizeau's experiment, 214, 256
flying saucer, 246
four-dimensional geometry, 71
fourth dimension, 18, 67, 72, 140, 160, 163, 169, 170, 191, 231, 242, 255, 298, 305
fraud, 9, 157, 161, 172, 183, 208, 214, 218, 220, 221, 222, 223, 224, 227, 231, 255
Fresnel, 14, 151, 210, 241, 251, 294, 297
Frisch, 173

—G—

Gale, 227
gamma rays, 44, 49, 50, 51, 58, 82, 83, 87, 252, 254, 307
Gehrcke, 275
Geiger counter, 272
Geitel, 268
General relativity, 68
General Theory of Relativity, 67, 73, 182, 183, 227, 241
Germer, 46
Giotto, 98
Goudsmit, 78
gravitational field, 64, 65, 68, 69
gravitational force, 8, 48, 53, 57, 58, 59, 61, 65, 67, 71, 119, 122, 124, 125, 133, 254
gravity, 8, 12, 23, 35, 52, 53, 54, 57, 58, 59, 60, 61, 63, 64, 65, 66, 67, 68, 70, 71, 72, 73, 76, 78, 87, 88, 97, 117, 118, 120, 127, 183, 226, 227, 231, 241, 242, 243, 254, 271
Great Comet, 93, 103

—H—

Hafele, 234
Hahn, 173
Halley's comet, 97, 98, 99, 100, 102
Hallwache, 268
Hargreaves, 183
heat, 25, 26, 58, 82, 87, 89, 98, 117, 118, 120, 121, 122, 123, 124, 125, 135, 136, 163, 172, 173, 181, 225, 252, 269, 270, 284
helium, 117, 119
Hertz, 14, 15, 151, 210, 211, 221, 251, 268, 297
high potential, 80
Hooke's, 13
Hubbard, 103
hulahoop, 106
hydrogen, 22, 36, 37, 78, 117, 118, 119, 120, 121, 122, 174, 251

—I—

inertia, 77, 165, 207, 242, 249
interference, 14, 29, 46, 227, 247, 252, 262, 275, 294
interference of light, 29, 262
interplanetary travel, 257
invention, 245, 250, 256
Ives, 166, 218

—J—

Jupiter, 13, 104, 119

—K—

Kantor, 229
Kaufmann, 223
Keating, 234
Kirchhoff, 14

—L—

Larmor, 17, 184
Leonard, 268
lightning, 30, 121, 122, 186, 252, 292
liquid, 38, 122, 123, 124, 125, 253
Lord Kelvin, 167, 251
Lorenz, 16, 17, 19, 73, 160, 163, 167, 168, 169, 184, 185, 186, 190, 191, 199, 206, 218, 220, 222, 223, 234, 235, 253
lower potential, 80

luminiferous ether, 198, 204, 211
Lummer, 275
Lynch, 206

—M—

M.A.U.D. Committee, 177
magnet, 8, 31, 32, 35, 38, 40, 41, 77, 88, 197, 199, 200, 201, 212, 213, 280, 281, 282, 290, 294
magnetic field, 8, 15, 19, 20, 31, 32, 35, 38, 39, 40, 65, 77, 84, 91, 101, 111, 112, 115, 125, 129, 130, 137, 141, 154, 155, 156, 201, 212, 213, 234, 280, 281, 282, 284, 286, 287, 288, 289
magnetic fields, 38, 82, 101, 208, 282
magnetic pressure, 16, 25, 115, 190, 234
magnetic storm, 34, 115, 122, 125
magnetic wind, 34, 156
magnetosphere, 113, 137, 138, 139, 153, 154, 226, 254
Manhattan project, 175, 176, 177, 256
mass increase, 17, 18, 19, 164, 186, 191, 215, 216, 221, 222, 223, 253
Maxwell, 14, 19, 21, 31, 89, 90, 151, 197, 198, 205, 207, 208, 210, 216, 221, 224, 240, 251, 266, 268, 274, 279, 280, 283, 284, 285, 286, 287, 288, 289
McCausland, 228
mechanical view, 218, 222, 297, 298
Melville, 13
meteors, 99, 138
Michelson, 15, 16, 20, 35, 137, 140, 150, 151, 152, 154, 156, 157, 184, 192, 199, 201, 202, 203, 214, 217, 226, 227, 255
Microwaves, 82
Minkowski, 18, 163, 170, 183, 191, 230, 231, 233, 242, 255
moon, 36, 37, 44, 48, 79, 103, 104, 105, 106, 107, 109, 253
Morley, 35, 140, 150, 151, 152, 155, 156, 157, 192, 202, 217, 226, 255

—N—

National Gallery, 236
negative charge, 78
neutron star, 132
neutron stars', 132
Newton, 13, 18, 65, 171, 180, 243, 262
nitrogen, 119
nonluminous spheres, 83
nuclear explosion, 24, 25, 26, 30, 34, 135, 136, 252
nuclear reactions, 8, 117, 120, 121, 124, 253

nucleus, 8, 22, 25, 26, 36, 37, 39, 40, 43, 44, 47, 49, 50, 51, 52, 53, 56, 57, 58, 59, 60, 63, 66, 67, 70, 71, 78, 80, 81, 82, 83, 87, 88, 89, 92, 96, 97, 98, 99, 100, 101, 102, 111, 133, 136, 138, 225, 227, 252, 254, 272

—O—

Ocean, 124, 249

—P—

Painleve, 76, 187
Painlevé, 187
painting, 236, 238, 239
pair production, 38, 44
particle accelerators, 92
photoelectric effect, 18, 19, 44, 90, 181, 221, 256, 266, 267, 268, 273, 275, 278
photon theory, 20, 160, 163, 171, 172, 209, 223, 224, 225, 228, 240, 262, 266, 267, 274, 278, 279
Planck, 16, 20, 142, 153, 166, 171, 172, 181, 182, 183
planetary rings, 8, 103
planets, 9, 12, 36, 53, 61, 79, 81, 103, 104, 132, 243, 245, 252, 254, 298
Poincare, 17, 166, 170, 183, 185
Poincaré, 163, 164, 165, 167, 168, 169, 170, 175, 185, 222
Poor, 195, 214, 243
positive charge, 78, 268
positron, 8, 38, 44, 45, 251
President Roosevelt, 175
principle of equivalence, 68, 71, 163, 182
propulsive force, 246, 247
propulsive panels, 247
protons, 22
Pulsars, 132

—R—

radio waves, 82, 84, 85, 121, 122, 125, 132, 133, 210, 211, 225, 252, 256
radioactive element, 25
radium, 25, 50, 51, 52
red shift effect, 108
reflection of light, 28, 224, 260, 275
reflection of sound, 259, 260
refraction of light, 28
refraction of sound, 261
relativity, 17, 18, 19, 20, 25, 68, 69, 70, 71, 73, 76, 87, 133, 142, 143, 147, 148, 149, 150, 154, 155, 156, 158, 159, 160, 161, 162, 163, 166, 167, 168, 169, 170,

172, 182, 183, 184, 185, 187, 191, 192, 194, 196, 198, 200, 204, 205, 206, 208, 212, 214, 215, 216, 217, 218, 219, 220, 221, 222, 225, 226, 227, 228, 229, 230, 231, 233, 234, 235, 239, 240, 241, 243, 253, 255, 264, 266, 267, 268, 271
resonance, 270
ringlets, 109
Ritz, 171
Robert Hooke, 13
Roemer, 143
Rudakov, 74, 164, 218, 235
Rutherford, 52, 157, 195, 217

—S—

Sachs, 175
Sagnac, 227
samareh, 8, 36, 37, 38, 39, 40, 41, 43, 44, 45, 47, 48, 78, 79, 82, 83, 84, 87, 88, 251, 252, 271
Santilli, 230
satellite, 106, 107, 109, 114, 119, 249
satellites, 13, 109
Saturn, 104, 106, 107, 108, 109
science fiction, 231
sea, 23, 122, 123, 125, 155, 156, 251
shock wave, 26, 46, 93, 98, 135, 136, 274
shock waves, 25, 26, 46, 93, 100, 102, 111, 112, 114, 136
Soddy, 217
solar flare, 125
solar flares, 35, 126, 156
solar system, 36, 48, 52, 79, 117, 127, 128, 129, 132
solar wind, 16, 20, 96, 97, 101, 111, 112, 113, 115, 127, 129, 130, 139, 140, 141, 153, 154, 203, 226, 254, 255, 298
Soldner, 171, 180
spaceship, 245, 246, 247, 248, 249
spark, 30, 91, 268, 277
sparks, 30, 252
spin of a star, 132
spin of electrons, 78
spin of the nucleus, 49, 50, 52, 53, 56, 66, 67, 71, 78, 83, 87, 88, 227, 254
spiral motion, 39, 40, 43, 79, 87, 88, 127, 252, 298
stars, 9, 12, 13, 50, 61, 83, 114, 118, 130, 132, 133, 142, 148, 180, 192, 221, 260, 263, 265, 271
stellar aberration, 19, 141, 142, 143, 148, 149, 202, 214, 221, 255
Stephen Hawking, 69
Stilwell, 218
Stoke, 15, 20, 142, 150, 152, 153, 214
Strassmann, 173

sun, 8, 13, 15, 36, 37, 48, 49, 52, 79, 81, 96, 98, 100, 101, 102, 105, 113, 117, 118, 119, 120, 121, 122, 123, 124, 125, 127, 128, 129, 130, 132, 141, 151, 180, 192, 221, 231, 253, 254, 261, 271, 273, 298
sunspots, 124
swindle, 217
Synchrotrons, 92
Szilard, 174

—T—

tail of a comet, 93, 96, 101, 252
tensor equations, 68
the statistical-kinetic theory of heat, 173
Thomson, 78
time dilation, 218
time machine, 169
Time Travel, 232
top, 60
Toronto Star, 236
Tunguska River, 135

—U—

Uhlenbeck, 78
unified field theory, 87
Unitary Field, 87
uranium, 25, 173, 174, 176, 177
Uranus, 103

—V—

Vega 1, 97

—W—

Wallace, 243
wave and particle theory, 239
weight, 23, 62, 70, 177, 242, 247
Whittaker, 158, 164, 167, 168, 172, 173, 182, 228, 275
Willard Gibbs, 173

—X—

x, 18, 39, 69, 78, 91, 158, 170, 171, 173, 208, 209, 211, 214, 219, 220, 223, 224, 225, 227, 241, 263, 267, 268, 269, 271, 272, 273, 275, 276, 277, 279, 282
x-rays, 30, 82, 83, 92, 225

—Y—

Young, 14, 210, 241, 251, 262, 294, 297

—Z—

Zahar, 167
Zinn, 174